MULTI-ROBOT SYSTEMS:
FROM SWARMS TO INTELLIGENT AUTOMATA

T0135043

MULTI-ROBOT SYSTEMS: FROM SWARMS TO INTELLIGENT AUTOMATA

Proceedings from the 2002 NRL
Workshop on Multi-Robot Systems

Edited by

Alan C. Schultz
Naval Research Laboratory,
Washington D.C., U.S.A.

and

Lynne E. Parker
Oak Ridge National Laboratory,
Oak Ridge, Tennessee, U.S.A.

KLUWER ACADEMIC PUBLISHERS
DORDRECHT / BOSTON / LONDON

A C.I.P. Catalogue record for this book is available from the Library of Congress.

ISBN 978-90-481-6046-4

Published by Kluwer Academic Publishers,
P.O. Box 17, 3300 AA Dordrecht, The Netherlands.

Sold and distributed in North, Central and South America
by Kluwer Academic Publishers,
101 Philip Drive, Norwell, MA 02061, U.S.A.

In all other countries, sold and distributed
by Kluwer Academic Publishers,
P.O. Box 322, 3300 AH Dordrecht, The Netherlands.

Printed on acid-free paper

Printed in the Netherlands.

Contents

Preface

In March 2002, the Naval Research Laboratory brought together leading researchers and government sponsors for a three-day workshop in Washington, D.C. on Multi-Robot Systems.

The workshop began with presentations by various government program managers describing application areas and programs with an interest in multi-robot systems. Government representatives were on hand from the Office of Naval Research, the Air Force, the Army Research Lab, the National Aeronautics and Space Administration, and the Defense Advanced Research Projects Agency.

Top researchers then presented their current activities in the areas of multi-robot systems and human-robot interaction. The first two days of the workshop concentrated on multi-robot control issues, including the topics of localization, mapping, and navigation; distributed surveillance; manipulation; coordination and formations; and sensors and hardware. The third day was focused on human interactions with multi-robot teams. All presentations were given in a single-track workshop format. This proceedings documents the work presented by these researchers at the workshop.

The invited presentations were followed by panel discussions, in which all participants interacted to highlight the challenges of this field and to develop possible solutions. In addition to the invited research talks, students were given an opportunity to present their work at poster sessions.

This workshop was held in advance of the formal meeting of the NATO working group IST-032/RTG-014 on Multi-Robot Systems in Military Domains. The workshop itself was held, in part, as a way to let the NATO working group members learn more about current efforts within the United States.

We would like to thank the Naval Research Laboratory for sponsoring this workshop and providing the facilities for these meetings to take place, and to the Office of Naval Research for their generous student travel grants.

We are extremely grateful to Magdalena Bugajska and Mitchell A. Potter for their vital help (and long hours) in editing these proceedings. Michelle Caccivio provided the administrative support to the workshop.

ALAN C. SCHULTZ AND LYNNE E. PARKER

I

LOCALIZATION, MAPPING AND NAVIGATION

ON THE POSITIONAL UNCERTAINTY
OF MULTI-ROBOT COOPERATIVE
LOCALIZATION

Ioannis M. Rekleitis, Gregory Dudek
Centre for Intelligent Machines, McGill University, Montreal, Québec, Canada
{ yiannis,dudek } @cim.mcgill.ca

Evangelos E. Milios
Faculty of Computer Science, Dalhousie University, Halifax, Nova Scotia, Canada
eem@cs.dal.ca

Abstract   This paper deals with terrain mapping and position estimation using multiple
robots. Here we will discuss work where a larger group of robots can mutually
estimate one another's position (in 2D or 3D) and uncertainty using a sample-
based (particle filter) model of uncertainty. Our prior work has dealt with a pair
of robots that estimate one another's position using visual tracking and coordinated motion and we extend these results and consider a richer set of sensing and
motion options. In particular, we focus on issues related to confidence estimation
for groups of more than two robots.

Keywords: Cooperative Localization, Multi-Robot Navigation, Position Estimation, Localization, Mapping.

1.   INTRODUCTION

In this paper we discuss the benefits of *cooperative localization* for a team
of mobile robots. The term *cooperative localization* describes the technique
whereby the members of a team of robots estimate one another's positions.
This is achieved by employing a special sensor (*robot tracker*) that estimates
a function of the pose of a moving robot relative to one or more stationary
ones (see section 1.1). Furthermore, we consider the effects of different robot
tracker sensors on the accuracy of localization for a moving robot *using only*
the information from the rest of the robots (as opposed to observations of the
environment). This approach results in an open loop estimate (with respect

A.C. Schultz and L.E. Parker (eds.), Multi-Robot Systems: From Swarms to Intelligent Automata, 3-10.
© 2002 *Kluwer Academic Publishers. Printed in the Netherlands.*

to the entire team) of the moving robot's pose without dependence on information from the environment. The experimental results allows us to examine the effectiveness of cooperative localization and estimate upper bounds on the error accumulation for different sensing modalities.

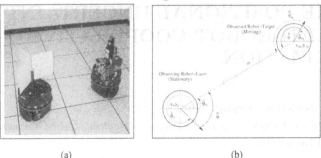

(a)   (b)

Figure 1.   (a) Two robots, one equipped with laser range finder (right) and the other with a target (left), employing cooperative localization. (b) Pose Estimation via Robot Tracker: Observation of the Moving Robot by the Stationary Robot. Note that the "camera" indicates the robot with the *Robot Tracker*; and $\hat{\theta}_w$, $\hat{\phi}_w$ are angles in world coordinates.

1.1   Cooperative Localization

Several different sensors have been employed for the estimation of the pose of one robot with respect to another robot. We restrict our attention to robot tracker sensors which return information in the frame of reference of the observing robot (i.e they estimate pose parameters relative to the robot making the observation). Consequently, for "two-dimensional robots" in a two dimensional environment, or for robots whose pose can be approximated as a combination of 2D position and an orientation, we can express the pose using three measurements; for ease of reference we represent them by the triplet $T = [\rho \; \phi \; \theta]$, where ρ is the distance between the two robots, ϕ is the angle at which the observing robot sees the observed robot relative to the heading of the observing robot, and θ is the heading of the observed robot as measured by the observing robot relative to the heading of the observing robot. (Figure 1b). If the stationary robot is equipped with the Robot Tracker, where $\mathbf{X}_m = [x_m, y_m, \theta_m]^T$ is the pose of the moving robot and $\mathbf{X}_s = [x_s, y_s, \theta_s]^T$ is the pose of the stationary robot then equation 1 returns the sensor output T:

$$\begin{bmatrix} \rho \\ \theta \\ \phi \end{bmatrix} = \begin{bmatrix} \sqrt{dx^2 + dy^2} \\ atan2(dy, dx) - \theta_s \\ atan2(-dy, -dx) - \theta_m \end{bmatrix}, \begin{array}{l} \text{Where}: \\ dx = x_m - x_s \\ dy = y_m - y_s \end{array} \qquad (1)$$

In order to estimate the probability distribution function (pdf) of the pose of the moving robot i at time t ($P(\mathbf{X}_i^t)$) we employ a particle filter (Monte Carlo simulation approach: see (Jensfelt et al., 2000; Dellaert et al., 1999; Liu et al., 2001)). The weights of the particles (W_i^t) at time t are updated using a Gaussian distribution (see equation 2 where $[\rho, \theta_i, \phi_i]^T$ has been calculated as in equation 1 but using the pose of particle "i" $(\mathbf{X}_{m_i})$ instead of the moving robot pose $(\mathbf{X}_m)$).

$$W_i^t = W_i^{t-1} \frac{1}{\sqrt{2\pi}\sigma_\rho} e^{\frac{-(\rho-\rho_i)^2}{2\sigma_\rho^2}} \frac{1}{\sqrt{2\pi}\sigma_\theta} e^{\frac{-(\theta-\theta_i)^2}{2\sigma_\theta^2}} \frac{1}{\sqrt{2\pi}\sigma_\phi} e^{\frac{-(\phi-\phi_i)^2}{2\sigma_\phi^2}} \quad (2)$$

The rest of the paper is structured as follows. The next Section 2 presents some background work. Section 3 contains an analysis and experimental study of the primary different classes of sensory information that can be naturally used in cooperative localization. Finally, Section 4 presents our conclusions and a brief discussion of future work.

2. PREVIOUS WORK

Prior work on multiple robots has considered collaborative strategies when the lack of landmarks made it impossible otherwise (Dudek et al., 1996). A number of authors have considered pragmatic multi-robot map-making. Several existing approaches operate in the sonar domain, where it is relatively straightforward to transform observations from a given position to the frame of reference of the other observers thereby exploiting structural relationships in the data (Leonard and Durrant-Whyte, 1991; Fox et al., 1998; Burgard et al., 2000). One approach to the fusion of such data is through the use of Kalman Filtering and its extensions (Roumeliotis and Bekey, 2000b; Roumeliotis and Bekey, 2000a).

In other work, Rekleitis, Dudek and Milios have demonstrated the utility of introducing a second robot to aid in the tracking of the exploratory robot's position (Rekleitis et al., 2000). In that work, the robots exchange roles from time to time during exploration thus serving to minimize the accumulation of odometry error. The authors refer to this procedure as *cooperative localization.*

Recently, several authors have considered using a team of mobile robots in order to localize using each other. A variety of alternative sensors has been considered. For example, (Kato et al., 1999) use robots equipped with omnidirectional vision cameras in order to identify and localize each other. In contrast, (Davison and Kita, 2000) use a pair of robots, one equipped with an active stereo vision and one with active lighting to localize. The various methods employed for localization use different sensors with different levels of accuracy; some are able to estimate accurately the distance between the robots, others the orientation (azimuth) of the observed robot relative to the observing robot and some are able to estimate even the orientation of the observed robot.

3.   SENSING MODALITIES

As noted above, several simple sensing configurations for a robot tracker are available. For example, simple schemes using a camera allow one robot to observe the other and provide different kinds of positional constraint such as the distance between two robots and the relative orientations. Moreover the group size affects the accuracy of the localization.

In the next part we present the effect the group size has on the accuracy of the localization for different sensors. The experimental arrangement of the robots is simulated and is consistent across all the sensing configurations. The robots start in a single line and they move abreast one at a time, first in ascending order and then in descending order for a set number of exchanges. The selected robot moves for 5 steps and after each step cooperative localization is employed and the pose of the moving robot is estimated. Each step is a forward translation by 100cm. Figure 2a presents a group of three robots, after the first robot has finished the five steps and the second robot performs the fifth step.

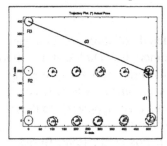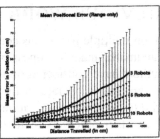

(a)   (b)

Figure 2.   (a) Estimation of the pose of robot R2 using only the distance from robot R1 (d1) and from robot R3 (d3). (b) Average error in position estimation using the distance between the robots only (3,4 and 10 robots; bars indicate std. deviation).

3.1   Range Only

One simple method is to return the relative distance between the robots. Such a method has been employed by (Grabowski and Khosla, 2001) in the millibots project where an ultra-sound wave was used in order to recover the relative distance. In order to recover the position of one moving robot in the frame of reference of another, at least two stationary robots (that are not collinear with the moving one) are needed thus the minimum size of the group using this scheme is three robots.

Estimating the distance between two robots is very robust and relatively easy. In experimental simulations, the distance between every pair of robots was estimated and Gaussian, zero mean, noise was added with $\sigma_p = 2cm$ regardless the distance between the two robots. Figure 2b presents the mean error per unit distance traveled for all robots, averaged over 20 trials. As can be seen in Figure 2b with five robots, the positional accuracy is acceptable with an error of 20cm after 40m traveled; for ten robots the accuracy of the localization is very good.

3.2 Azimuth (Angle) Only

Several robotic systems employ an omnidirectional vision sensor that reports the angle at which another robot is seen. This is also consistent with information available from several types of observing systems based on pan-tilt units. In such cases orientation at which the moving robot is seen can be recovered with high accuracy. We performed a series of trials using only the angle at which one robot is observed, using groups of robots of different sizes. As can be seen in Figure 3 the accuracy of the localization does not im-

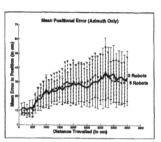

Figure 3: Average error in position estimation using the orientation of the moving robot is seen by the stationary ones.,

prove as the group size increases. This is not surprising because small errors in the estimated orientation of the stationary robots scale non-linearly with the distance. Thus after a few exchanges the error in the pose estimation is dominated by the error in the orientation of the stationary robots.

To illustrate the implementation of the particle filter, we present the probability distribution function (pdf) of the pose of the moving robot after one step (see Figure 4). The robot group size is three and it is the middle robot R2 that moves. The predicted pdf after a forward step can be seen in the first subfigure (4a) using odometry information only; the next two subfigures (4b,4c) present the pdf updated using the orientation at which the moving robot is seen by a stationary one (first by robot R1 then by robot R3); finally, the subfigure 4d presents the final pdf which combines the information from odometry and the observations from the two stationary robots. Clearly the uncertainty of the robot's position is reduced with additional observations.

3.3 Position Only

Another common approach is to use the position of one robot computed in the frame of reference of another (relative position). This scheme has been employed with two robots (Burgard et al., 2000) in order to reduce the uncertainty. The range and azimuth information ($[\rho, \theta]$) is combined in order to improve the pose estimation. As can be seen in Figure 5a even with three robots the error in pose estimation is relatively small (average error 30cm for 40m distance traveled per robot, or 0.75%). In our experiments the distance between the two robots was estimated and, as above, zero-mean Gaussian noise was added both to distance and to orientation with $\sigma_\rho = 2cm$ and $\sigma_\theta = 0.5°$ respectively.

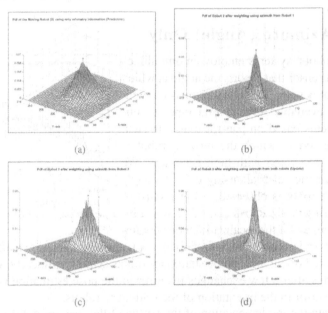

(a)   (b)

(c)   (d)

Figure 4.   The pdf of the moving robot (R2) at different phases of its estimation: (a) prediction using odometry only; (b) using the orientation from stationary robot R1; (c) using the orientation from stationary robot R3; (d) final pdf.

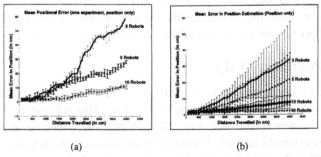

(a)   (b)

Figure 5.   Average error in position estimation using both the distance between the robots and the orientation the moving robot is seen by the stationary ones. (a) Average error in positioning of the team of robots one trial (3,5 and 10 robots). (b) Average error in position estimation over twenty trials (3,5, 10 and 40 robots).

The experiment was repeated twenty times and the average error in position is shown in Figure 5b for groups of robots of size 3,5,10 and 40.

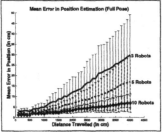

# of Robots	3	5	10
Range (ρ)	38.80	21.63	8.13
Azimuth (θ)	32.83	32.20	
Position (ρ, θ)	34.25	21.79	7.50
Full Pose (ρ, θ, ϕ)	28.73	16.72	6.08

Figure 6. Average error in position estimation using full pose $[\rho, \theta, \phi]$.

Table 1. The mean error in position estimation after 40m travel over 20 trials.

3.4   Full Pose

Some robot tracker sensors provide accurate information for all three parameters $[\rho, \theta, \phi]$ and they can be used to accurate estimate the full pose of the moving robots (Kurazume and Hirose, 1998; Rekleitis et al., 2001). In the experimental setup the robot tracker sensor was characterized by Gaussian, zero mean, noise with $\sigma = [2cm, 0.5°, 1°]$. By using the full equation 2 we weighted the pdf of the pose of the moving robot and performed a series of experiments for 3, 5 and 10 robots; with very low positional error (see Figure 6).

4.   CONCLUSIONS

In this work we examined the effect of the size of the team of robots and the sensing paradigm on cooperative localization (see Table 1 for a synopsis). Also, preliminary results from experiments with varying odometry error have shown that cooperative localization is robust even with 10-20% odometry errors.

In future work we hope to further extend the uncertainty study for different group configurations and motion strategies. An interesting extension would be for the robots to autonomously develop a collaborative strategy to improve the accuracy of localization (Potter et al., 2001). Given a large group of robots, an estimate of the effects of team size on error accumulation would allow the group of be effectively partitioned to accomplish sub-tasks while retaining a desired level of accuracy in positioning.

References

Burgard, W., Fox, D., Moors, M., Simmons, R., and Thrun, S. (2000). Collaborative multi-robot exploration. In *Int. Conf. on Robotics & Automation*, p. 476–481.

Davison, A. and Kita, N. (2000). Active visual localisation for cooperating inspection robots. In *IEEE/RSJ Int. Conf. on Intelligent Robots & Systems*, p. 1709–1715.

Dellaert, F., Burgard, W., Fox, D., and Thrun, S. (1999). Using the condensation algorithm for robust, vision-based mobile robot localization. In *IEEE Computer Society Conf. on Computer Vision & Pattern Recognition*. IEEE Press.

Dudek, G., Jenkin, M., Milios, E., and Wilkes, D. (1996). A taxonomy for multiagent robotics. *Autonomous Robots*, 3:375–397.

Fox, D., Burgard, W., and Thrun, S. (1998). Active markov localization for mobile robots. *Robotics & Autonomous Systems*. To appear.

Grabowski, R. and Khosla, P. (2001). Localization techniques for a team of small robots. In *IEEE/RSJ Int. Conf. on Intelligent Robots & Systems*, p. 1067–1072.

Jensfelt, P., Wijk, O., Austin, D., and Andersso, M. (2000). Feature based condensation for mobile robot localization. In *Int. Conf. on Robotics & Automation*.

Kato, K., Ishiguro, H., and Barth, M. (1999). Identifying and localizing robots in a multi-robot system environment. In *Int. Conf. on Intelligent Robots & Systems*.

Kurazume, R. and Hirose, S. (1998). Study on cooperative positioning system - optimum moving strategies for cps-iii. In *Proc. IEEE Int. Conf. on Robotics & Automation*, volume 4, p. 2896–2903.

Leonard, J. J. and Durrant-Whyte, H. F. (1991). Mobile robot localization by tracking geometric beacons. *IEEE Transactions on Robotics & Automation*, 7(3):376–382.

Liu, J. S., Chen, R., and Logvinenko, T. (2001). A theoretical framework for sequential importance sampling and resampling. *Sequential Monte Carlo in Practice*.

Potter, M. A., Meeden, L. A., and Schultz, A. C. (2001). Heterogeneity in the coevolved behaviors of mobile robots: The emergence of specialists. In *Seventeenth Int. Joint Conf. on Artificial Intelligence (IJCAI)*.

Rekleitis, I. M., Dudek, G., and Milios, E. (2000). Multi-robot collaboration for robust exploration. In *Proc. of Int. Conf. in Robotics & Automation*, p. 3164–3169.

Rekleitis, I. M., Dudek, G., and Milios, E. (2001). Multi-robot collaboration for robust exploration. *Annals of Mathematics & Artificial Intelligence*, 31(1-4):7–40.

Roumeliotis, S. I. and Bekey, G. A. (2000a). Bayesian estimation and kalman filtering: A unified framework for mobile robot localization. In *Int. Conf. on Robotics & Automation*, p. 2985–2992.

Roumeliotis, S. I. and Bekey, G. A. (2000b). Collective localization: A distributed kalman filter approach to localization of groups of mobile robots. In *Int. Conf. on Robotics & Automation*, p. 2958–2965.

A MULTI-AGENT SYSTEM FOR MULTI-ROBOT MAPPING AND EXPLORATION

Kurt Konolige[1]
SRI International

Didier Guzzoni
VerticalNet

Keith Nicewarner
Nasa Ames Research Center

Abstract: Multi-robot mapping and exploration involves a team of mobile robots cooperating autonomously to discover information about an area, construct a coherent geometric map of the area, and find other relevant objects in the area, e.g., people or machinery. Additionally, the team must interact with humans, both to send them useful information, and to accept commands influence their mission. Making the physical robots into agents, and connecting them with other software agents for control and communication, enables us to dynamically configure and task a complex system.

Keywords: Autonomous robots, multi-agent teams, mapping

1.   INTRODUCTION

In SRI's Sense Net project (Konolige *et al.,* 1999), we have configured a set of mobile robots (mobots) as a cooperative team that is tasked for rescue,

[1] The Sense Net project work is performed as part of DARPA's Tactical Mobile Robot initiative, under Contract # DAAE07-98-C-L030.

A.C. Schultz and L.E. Parker (eds.), Multi-Robot Systems: From Swarms to Intelligent Automata, 11-19.
© 2002 *Kluwer Academic Publishers. Printed in the Netherlands.*

hazard detection, and surveillance in urban settings. The mobots survey an indoor area of interest, and transmit an annotated map back to their controllers, highlighting hazards and human and mechanical activity.

Our approach is to consider each mobot as an autonomous unit, with a sufficient complement of sensors (e.g., visual stereo, laser range finder) to recover the 3D geometry of a scene for local mapping. While individual, autonomous mobots are the basis of the Sense Net, mapping and surveillance of a large area can be performed more quickly and reliably by using several mobots working collaboratively, some with specialized capabilities. The Sense Net is built up through the coordination of mobot activity, as the mobots communicate over wireless links to integrate local information into a more global picture of the task area, which can be relayed to a base station, or shared among mobots.

From a systems perspective, the challenge is to coordinate complex individual mobots, which have some degree of autonomy, with each other and with a set of software programs that implement higher-level interpretation, presentation, and control functions. If we demand that each mobot be connected to an operator station, where its internal functions are monitored and controlled, then the amount of operator attention needed to control the mobots becomes unmanageable. Instead, we would like a single operator to be able to effectively understand and control the mobot team (other operators and observers should also be accommodated, of course). To do so, we take advantage of agent technology to dynamically construct a large, cooperative system consisting of the mobots, associated programs, and human operators.

Some of the issues that have arisen in the construction of the Sense Net as an agent community are the following.

- **Naming and connection/reconnection services.** The mobots operate on a wireless ethernet, and the link is unreliable. They must be able to operate autonomously in the absence of outside links, but also be able to reconnect and find each other when the link is established.

- **Independent agents.** Each mobot can have a set of sensors, some of which are tightly integrated with navigation capabilities (e.g., the laser range finder), and some that are more independent (e.g., a panoramic camera). Rather than view a mobot as a single agent, it is useful to see it as a collection of agents, which can be dynamically configured and controlled by other agents.

- **Information repository.** It is inefficient for an operator or mobot to send queries to a set of mobots to find out their positions, since the number of such queries will grow quadratically with the number of mobots. Instead, often-requested information can be cached in a central repository, and kept updated by each mobot, so that a snapshot of the system is available with a single request.

- **Information broadcast.** In distributed mapping, all mobots are constantly trying to incorporated map information from other mobots into their own maps. Such information is most efficiently transmitted by broadcast methods. Here the challenge is to make sure that the mapping process converges if the mobots are connected often enough.

- **High bandwidth links.** For video or picture data coming from the mobots to an operator, a high-speed connection that bypasses normal agent query-response protocols is necessary. In this case, the agent interface can mediate the connection and reconnection of such peer-to-peer links.

- **Distributed presentation and control.** Operators must be able to understand what the Sense Net is doing, and dynamically access and control services provided by the Net. Presentation agents dynamically access information coming from the mobots, and construct a coherent geometric view of the distributed mapping process, including placement of mobots, people-tracking, panoramic video, and other services. Additionally, interface agents allow the operator to interact with the Net in easily-learned ways, such as pen gestures to indicate where the mobots should explore or set up surveillance.

The Sense Net is an implemented and working system, utilizing the agent technology provided by SRI's Open Agent Architecture (OAA) (Moran *et al.*, 1996, 1997). In the rest of the paper, we describe the Sense Net system, concentrating on the use of agents to provide a coherent framework for distributed mapping and control.

2.   MOBOT MAPPING CAPABILITIES

Exploration of an unknown interior or exterior area, and robust localization within the mapped area, are critical capabilities for a mobot. Our current research suggests that effective use of range information to create local geometric maps is possible, using processing available on small mobots. The result is that each mobot is able to create a rich, 3D model of its environment, as it moves, in real time.

Geometry-based methods for mapping rely on the fusion of local range sensing to build up a model or map of the local environment. This model is then used by the mobot to self-localize as it explores. Recent methods have efficiently solved this problem in well-structured indoor environments, using data from a laser range-finder (Burgard *et al.*, 1999; Gutmann and Konolige, 1999).

In our approach, based on work by Lu and Milios (Lu and Milios, 1997), the mobot's positions are treated as a set of random variables, with overlapped range readings and dead reckoning forming local constraints between the positions. We introduce two techniques to implement a complete map-building system, namely, Local Registration for efficiently adding new data to the map, and Global Correlation for finding and closing large cycles.

Local Registration takes the most recent range readings, usually over a distance of several meters, and registers them using the method of Lu and Milios. We have found that, for most environments, there is an optimal fusing distance for local registration. Readings fused over a shorter distance will not have enough information to be consistently integrated, while those over longer distances add more computational work without generating better results.

To determine the topological relationship between poses that close a cycle, we compare a recent portion of the map around the current pose with the older portions of the map. Where there is a good match, it is likely that the new pose is topologically connected to one of the older poses. In our current method of global registration, once a topological connection is made, it is not possible to undo it, since all poses are updated and no history is kept. Therefore, any such connection needs to be very certain before it is made, and false positive rejection is critical. This is the main reason for matching a patch that integrates several scans, and also for providing post-match filters to reject false positives.

Another constraint on map matching is that it must be efficient, since we intend to run it constantly in the background as the robot starts cycling back to places previously visited. Recent investigations by one of the authors has provided a fast and accurate matching technique based on correlation (Konolige, 1999). The justification for this technique lies in a Bayesian analysis of the match probability. For any given new map patch r rand old map mm, we seek the posterior probability $p(l \mid r, m)$ that the robot is at pose l. Using Bayes' rule, we have:

$$p(l \mid r, m) = k \cdot p(r \mid l, m)p(l, m)$$

The sensor response function $p(r \mid l, m)$ is the probability that we would see the map patch from the robot pose l, given the old map. As shown in (Konolige, 1999), the sensor response can be approximated by a correlation operator. A regular grid is imposed on the map area, and for each cell we calculate the probability $p(r_i)$ of the map patch impinging on the cell and $p(m_i)$ of the old map impinging on the cell. The correlation operator is:

$$\sum_i p(r_i)p(m_i)$$

In practice it is convenient to put all the uncertainty into the map probability $p(m_i)$, simplifying the above sum and making it amenable to optimized implementation.

We have tested the LRGC algorithm in many different environments, using range scans from the SICK laser range-finder. The algorithm as currently implemented runs in real-time on a standard PC, with only slight pauses of a few seconds when a large loop is closed.

3. DISTRIBUTED MAPPING

Once we have each mobile robot building a locally consistent map (as described in the previous section), we must then distribute the maps so that a globally consistent map can be built. A common method for this is to send the maps from each robot to a central global mapper. This design is appealing for several reasons, but a fully decentralized approach has benefits that in the long run outweigh the appeal of the centralized approach.

3.1 Centralized Mapping

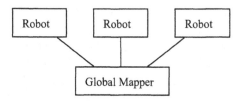

One reason to choose a centralized mapper design is simplicity: a centralized mapper is the logical arbiter of any discrepancies. Another reason is the economics of computational power. Until recently, the limited computational power available on most mobile robot platforms forced a centralized approach.

3.2   **Decentralized Mapping**

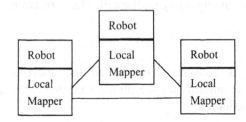

Instead of sending scans and updates to a central mapper, we choose to broadcast the information to any node that can hear. Nodes that are not in direct communication with each other can use other nodes as "routers." For example, a robot that is not currently in direct communication with a remote robot can use the robots in between to relay the information.

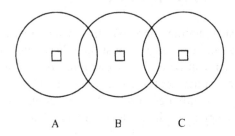

Figure 1. Robot B acts as router between A and C.

With a fully decentralized approach, there is no single point of failure. If one node of the network is lost, or a link fails, the rest of the network of robots can continue functioning. In the centralized approach, a loss of the central server node or links to it would be fatal. Decentralized mapping can display "graceful degradation" in the face of network faults. In fact, disconnected subnets of robots can form so that each robot has access to all available information (albeit a reduced set) to perform mapping.

A particularly appealing feature of the decentralized method is that proximate updates from nearby robots are received quickly, whereas remote updates from robots along multi-hop routes are delayed. Information is automatically prioritized by relevance, with most relevant information being readily available and least relevant information being delayed.

With a centralized approach, communication with the global mapper is critical. Thus, reliable communication strategies (such as TCP IP) are typically used. Unfortunately, TCP messages require handshaking that increase the burden on an already loaded communications channel. But with a decentralized approach, a best-effort communication strategy (such as UDP IP) can be used because getting the information to a particular network node is not as important as getting the information to every host that might be interested. In addition, UDP broadcasts take best advantage of the limited communication channel bandwidth since they don't require acknowledge replies.

Our decentralized broadcast strategy is as follows. For each robot:

1. Get new scan from sensor.
2. Broadcast new scan.
3. Register new scan with local map.
4. Broadcast updates to scans changed by registration.

Each robot maintains a separate list of scans for each robot it has heard from. The local map is correlated against maps from other robots until they can be localized with respect to each other. After that, the local registration procedure takes other robot maps into account, providing each robot with a locally resident globally consistent map.

4. THE AGENT INTERFACE AND OAA

It is our belief that cooperative problem-solving by mobots requires the use of flexible, high-level strategies that can be adapted by individual agents based on their progress and operating environment. This methodology contrasts with self-organizational methods, in which no global control policy exists. Rather, 'control' within self-organizing systems 'emerges' as a result of the actions of individual agents. Self-organizing approaches are not well-suited to tasks that require guaranteed accomplishments of specific tasks, as is the case for exploration and surveillance.

Our approach to cooperative problem-solving in the tactical mobile robotics domain is based on strategies that are declaratively encoded, including specifications of conditions that must be satisfied for them to apply, statements of their effectiveness under various conditions, and possible negative effects. Specification of this information will enable informed trade-off analyses to be performed when candidate strategies are evaluated for use in different situations. This analysis is founded on ideas from utility theory applied in the planning domain (Bratman, 1998). We explicitly do not want to implement an approach based on economic or negotiation models (Davis and Smith, 1983; Wellman, 1993, Rosenschein and Zlotkin, 1993). These models suffer from the drawback that convergence is generally not guaranteed, and even when it occurs may take a long time. As such, they are not well suited to the kinds of time-critical tasks envisioned for this project, especially since the dynamic environment in which the mobots operate will require frequent, rapid adjustments to the individual and multi-agent problem-solving strategies.

Our implementation of distributed mapping uses the dynamic connection services of OAA, and the reactive strategies of the Procedural Reasoning System (PRS), which provides an architecture for reactive multi-agent control (Konolige and Myers, 1998). The OAA has a distributed architecture in which a Facilitator agent is responsible for scheduling and maintaining the flow of communication among number of client agents. Each robot agent provides information about its cognitive and physical state and accepts commands to control the mobile platform it controls. The information includes position with respect to the robot's internal coordinate system, and map information from exploration.

During movement, each robot keeps track of its local map, as described in previous sections. It communicates with the database agent to update its position about once a second, and to report any new objects that it finds, so they can be incorporated into the global database and made available to other agents. In this way, the database agent has available information about all of the robot agents that are currently operating.

References

Bratman, M. E., Israel, D. J., and Pollack, M. E. (1988). Plans and resource-bounded practical reasoning. *Computational Intelligence*, 4(4).

Burgard, W., Fox, D., Jans, H., Matenar, C., and Thrun, S. (1999). Sonar-based mapping of large-scale mobile robot environments using EM. In *Proceedings of the International Conference on Machine Learning (ICML' 99)*.

Davis, R. and Smith, R. G. (1983). Negotiation as a metaphor for distributed problem solving. *Artificial Intelligence*, 20: 63-109.

Gutmann, J. S. and Konolige, K. (1999). Incremental Mapping of Large Cyclic Environments." In *Proceedings of CIRA 99*, Monterey, California.

Konolige, K. (1999). Markov localization using correlation. *In Proceedings International Joint Conference on Artificial Intelligence (IJCAI' 99)*, Stockholm, Sweden.

Konolige, K. and Myers, K. (1998). The SAPHIRA architecture: a design for autonomy. In *Artificial Intelligence Based Mobile Robots: Case Studies of Successful Robot Systems*, Kortenkamp, D., Bonasso, R. P., and Murphy, R., editors, MIT Press.

Konolige, K., Gutmann, S., Guzzoni, D., Ficklin, R., and Nicewarner, K. (1999). A Mobile Robot Sense Net. In *Proceedings of the SPIE*, Boston, MA.

Lu, F. and Milios, E. E. (1997). Globally consistent range scan alignment for environment mapping. *Autonomous Robots*, 4(4).

Moran, D. B., Cheyer, A. J., Julia, L. E., Martin, D.L., and Park, S. (1996). The Open Agent Architecture and Its Multimodal User Interface. SRI Tech Note.

Moran, D. B., Cheyer, A., Julia, L., Martin, D., and Park, S. K. (1997).. Multimodal User Interfaces in the Open Agent Architecture. In *Proceedings of IUI97*, Orlando, FL.

Rosenschein, J. S. and Zlotkin, G. (1993). A domain theory for task-oriented negotiation. In *Proceedings of the Thirteenth International Joint Conference on Artificial Intelligence*, AAAI Press.

Wellman, M. P. (1993). A market-oriented programming environment and its application to distributed multicommodity problems. *Journal of Artificial Intelligence Research*, 1:1-23.

DISTRIBUTED HETEROGENEOUS SENSING FOR OUTDOOR MULTI-ROBOT LOCALIZATION, MAPPING, AND PATH PLANNING

Lynne E. Parker, Kingsley Fregene*, Yi Guo, and Raj Madhavan
Computer Science and Mathematics Division
Oak Ridge National Laboratory, P.O. Box 2008, Oak Ridge, TN 37831-6355 †
parkerle@ornl.gov,guoy@ornl.gov,kocfrege@ieee.org,raj_madhavan@yahoo.com

Abstract   Our objective is to develop a team of autonomous mobile robots that are able to operate in previously unfamiliar outdoor environments. In these environments, the robot teams should be able to cooperatively localize even when DGPS is not consistently available, to autonomously generate rough elevation maps of their terrain, and to use these generated maps to plan multi-robot paths that enable them to accomplish their mission objective, such as reconnaissance and surveillance or perimeter security. This paper briefly outlines our approaches to achieving this objective, along with some of our implementation results on our team of four ATRV-mini mobile robots.

Keywords: Distributed sensing, heterogeneous robots, outdoor navigation.

1.   INTRODUCTION

In practical applications of teams of mobile robots in outdoor terrains, a serious consideration is the navigation of the robots across previously unfamiliar terrain. For nearly all applications, these robots must be able to move safely to avoid navigation hazards. However, for many applications *safe* navigation alone is not sufficient; the robots are also required to find *efficient* paths

---

*Research performed while author visited ORNL. Author's current address is University of Waterloo, Ontario, Canada.
†This research is sponsored in part by the Engineering Research Program of the Office of Basic Energy Sciences, U. S. Department of Energy. Accordingly, the U.S. Government retains a nonexclusive, royalty-free license to publish or reproduce the published form of this contribution, or allow others to do so, for U. S. Government purposes. Oak Ridge National Laboratory is managed by UT-Battelle, LLC for the U.S. Dept. of Energy under contract DE-AC05-00OR22725.

A.C. Schultz and L.E. Parker (eds.), Multi-Robot Systems: From Swarms to Intelligent Automata, 21-30.
© 2002 *Kluwer Academic Publishers. Printed in the Netherlands.*

through their terrain based upon their mission requirements. These robots may need to operate for a period of time in an outdoor area, and may need to develop knowledge about the outdoor terrain. For example, reconnaissance and surveillance tasks may require the robots to set up security patrols based upon terrain visibility. Exploration tasks may require robots to search an outdoor area for an object or feature of interest. Military applications may require a robot team to move from point to point within given boundaries along the most efficient route possible.

All of these practical applications require the robot teams to be able to 1) localize within the outdoor environment, 2) map their terrain sufficiently to enable efficient path planning, and 3) plan their paths according to the mission goals. A significant amount of research has addressed these individual and related problems, including localization, mapping for indoor planar environments, cross-country and road-following navigation, following trajectories roughly specified by a human operator, and path planning for both indoor and outdoor environments given a terrain map. In particular, the cooperative localization and mapping issue has been very extensively studied. However, most of this prior research has addressed the indoor environment. Very little prior research has addressed the complete problem of developing approaches that enable a team of robots to be immediately placed in a previously unfamiliar outdoor environment, to generate sufficient knowledge of the terrain for safe and efficient navigation, and to derive efficient multi-robot path plans.

A key challenge in this research is enabling the robot team to autonomously develop a terrain map of their outdoor working environment. Commonly available Digital Elevation Maps (DEMs) are not provided at the terrain resolution needed for safe and efficient robot navigation. The motion of a robot across a terrain using the Differential Global Positioning System (DGPS) to make position and elevation measurements will not operate in environments that include trees, buildings, steep hills, and so forth that generate a multi-pathing problem for DGPS. Even if continual DGPS could be guaranteed, robots would still need an additional mechanism for recognizing obstacles and unsafe navigation regions that should not be entered. In many applications, human operators could mark unsafe regions on an image roughly correlated to the DGPS positions, but this approach does not address the need to have a map model with sufficient elevation detail to enable efficient, repeated navigation across the working area.

An ideal solution would be to automate the challenging aspects of this problem so that the robot team can indeed be placed in a new, outdoor environment and operate successfully according to the mission requirements. Our research is aimed at developing the algorithms and the overall system that will enable this type of application to be solved with robot teams. Our approach takes advantage of the heterogeneous distributed sensing capabilities afforded by a team of multiple robots. Robots should be able to assist each other as needed to provide collaborative sensing capabilities that enable them to accomplish their mission.

The problem statement for the robot team that we are addressing is as follows: given an unknown outdoor environment with incomplete DGPS availability and unsafe navigation regions, develop an elevation map of the terrain marked with the unnavigable areas and use this map to plan multi-robot paths that satisfy the mission objectives (such as patrol paths). For the purposes of this research, the unsafe navigation regions are considered to be positive obstacles (e.g., trees, large rocks, etc.) and areas whose slope or local roughness exceeds a pre-specified limit. For now, we are not addressing the recognition of negative obstacles (e.g., holes in the ground or pools of water) or hidden obstacles (e.g., in grassy areas), since much recent work is addressing this issue and it is expected that these approaches can easily be inserted into our system.

The organization of this paper is as follows: Section 2 gives an overview of the experimental setup. The next three sections then outline the approaches to the three key issues in this research – multi-robot localization in Section 3, multi-robot mapping in Section 4, and multi-robot path planning in Section 5. Examples of the results of our implementation to date are given in Section 6. We conclude with summary remarks in Section 7.

2. ROBOT TEAM AND EXPERIMENTAL SETUP

The experimental platform (see Figure 1) is a team of four ATRV-Mini wheeled mobile robots with 4-wheel differential-drive skid-steering. The experimental setup consists of a wireless mini-LAN, a Local Area DGPS (LADGPS), a software platform (*Mobility* from RWI) and codes developed in-house under Linux to read and log the data for the sensors on each robot. The wireless LAN is set up outdoors between an Operator Console Unit (OCU) and the robots. The OCU consists of a rugged notebook equipped with a Breeze-COM access point and antennas. Each robot has a BreezeCOM station adapter and an antenna. The LADGPS is formed by the base station/antenna hardware connected to the OCU and remote stations/antennas directly mounted on each robot. Each robot's station receives differential corrections from the base station such that LADGPS accuracy of up to 10 centimeters is obtainable. The distributed CORBA-based interface offered by *Mobility* ensures that querying the sensor slots of particular robots is done in a transparent decentralized manner by simply appending the robot's ID to all such queries.

The sensor suite is comprised of encoders that measure the wheel speeds and heading, DGPS, and a magnetic compass. Two of the robots are equipped with

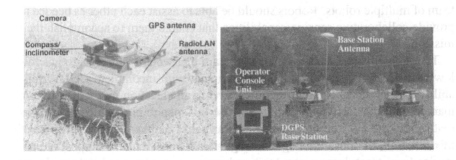

Figure 1. The ATRV-Mini sensor suite and experimental setup. The sensor suite consists of encoders, DGPS, a compass and a PTZ camera. The experimental setup depicted in the second photo consists of an operator console unit, a DGPS base station and a base station antenna. See text for further details.

a pan-tilt-zoom (PTZ) capable camera for visual perception and the remaining two robots are equipped with a SICK scanning laser rangefinder.

3.   DISTRIBUTED EKF LOCALIZATION

In outdoor environments, errors introduced due to distance traveled can be significant and unpredictable. This is a direct consequence of the undulatory nature of the terrain of travel and the uncertainties introduced into sensor data. These challenges make it comparatively difficult to realize successful navigation in unstructured outdoor environments. Motivated by these factors, our approach is an Extended Kalman Filter (EKF) based multi-robot heterogeneous localization framework similar to that developed in (Roumeliotis and Bekey, 2000), but differing in the following ways: 1) the kinematic model of the robots is nonlinear, 2) no absolute positioning system capable of providing relative pose information is assumed to be available, and 3) the robots traverse on uneven and unstructured outdoor terrain. In the first case, a kinematic model that sufficiently captures the nonlinear vehicle motion is key to efficient use of sensor data and is central to successful autonomous navigation. A nonholonomic robot with a nonlinear kinematic model performs significantly better as the model efficiently captures the maneuvers of the robot. In the second case, even though we consider systems including DGPS, it only provides absolute position information for a single robot subject to the number of satellites in view at any given time. DGPS is not guaranteed to be continually available.

When some robots of the team do not have absolute positioning capabilities or when the quality of the observations from the absolute positioning sensors deteriorate, another robot in the team with better positioning capability can assist in the localization of the robots whose sensors have deteriorated or failed.

In such cases, if relative pose information is obtained, casting the EKF-based localization algorithm in a form such that the update stage of the EKF utilizes this relative pose thereby provides reliable pose estimates for all the members of the team. Under this approach, at least one of the robots must maintain global positioning. (In future work, we will be examining how to maintain this constraint as the robots perform their primary mission.) We obtain relative pose information through one of two ways – a scanning laser range finder-based, and a vision-based cooperative localization approach.

In the case of cooperative localization via laser, consider the two-robot cases where robot #2 has a scanning laser range finder. The localization process proceeds as follows. First, robot #2 identifies robot #1 and acquires a range and bearing laser scan. Then, after the necessary preprocessing to discard readings that are greater than a predefined threshold, the range and bearing to the minima identified in the laser profile of robot #1 are determined. Finally, from the range and bearing pertaining to the minima, the pose of robot #2 is inferred and relative pose information is available for use.

In the case of vision-based cooperative localization, the robot's camera is used to provide relative position information. In the case where two robots are performing cooperative localization with the camera-equipped robot #1 lacking in absolute positioning capability, relative position information is obtained as follows. First, robot #1 searches the vicinity for another robot (say, robot #2) whose pose is known (this is determined via communication). Robot #1 then visually acquires robot #2 using an object recognition algorithm. The algorithm identifies the centroid of the robot within the image frame using a color segmentation scheme and marks its pixel coordinates on that frame. An incremental depth-from-motion algorithm (see (Fregene *et al.*, 2002) for more details) computes the depth for a window within the frame that encloses these coordinates. The required relative position is inferred from the computed depth and the bearing of robot #2 relative to robot #1 is approximately determined from the lateral displacement between the enclosed pixel coordinates and the coordinates of the frame's optical center. The robot states are then updated. More details on these approaches are available in (Madhavan *et al.*, 2002).

4. MULTI-ROBOT MAPPING

Incremental terrain mapping takes place via four main processes. An incremental dense depth-from-camera-motion algorithm (which is an adaptation of the work reported in (L. Matthies, *et al.*, 1989)) is used to obtain depth ranges to various features in the environment. The relative pose of the robots at these locations are associated with particular depth information. An elevation gradient of the terrain is determined by fusing GPS altitude information and vertical displacements obtained from inclinometer pitch angles. The depth

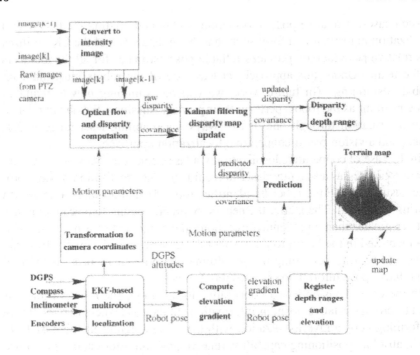

Figure 2.   The overall terrain mapping scheme

and elevation information are then registered with their associated covariances. The terrain map is updated to incorporate the registered values at their proper coordinates. The covariances associated with each measurement provides the confidence the algorithm has in that measurement. In the case of overlapping areas, this confidence determines whether or not the map is updated. The overall schematic diagram of the algorithm is shown in Figure 2. More details on our approach are available in (Fregene *et al.*, 2002).

5.   MULTI-ROBOT PATH PLANNING

Our multi-robot path planning approach operates as follows. First, each robot plans its own path independently using D^*. The path is broadcast to all other robots, so every robot knows all path information. Under our approach, the paths that are planned for each robot are fixed, i.e., the following steps will not alter the (x, y) sequences of the paths. Instead, we define velocity profiles so that, while robots follow their paths, they insert delays as required to avoid collisions. Once the paths are planned, the collision check is then executed. If the collision is a time-space collision, that is, two or more robots reach the same point at the same time, an N-dimensional coordination diagram (CD) is constructed with collision regions marked as obstacles in the diagram.

D^* searches for a free trajectory in the coordination diagram. The trajectory is then interpreted into a velocity profile for each robot, and the performance index of the current trajectory solution is calculated. Since the searching in CD is distributed across the robots, each search can take a different cost function to minimize based upon differences in priorities between robots at intersections. Then the performance index and velocity profile are broadcast to all other robots. An evaluation is done to get a minimum value of the performance index, and the corresponding velocity profile is chosen. More details on our approach are available in (Guo and Parker, 2002).

6.   EXPERIMENTAL RESULTS

We have implemented portions of this approach for multi-robot localization, mapping, and path planning in outdoor environments using distributed sensing. We briefly mention some of these results here, referring the reader to (Fregene *et al.*, 2002; Madhavan *et al.*, 2002; Guo and Parker, 2002) for more details.

Figures 3 and 4 show the results for the laser-based cooperative localization described in Section 3. Figure 3 shows the estimated paths of robots #1 and #2. The pose standard deviations of robot #2 in Figure 4 demonstrate the utility of the relative pose information in accomplishing cooperative localization. At $time = 21$ seconds, DGPS becomes unavailable as indicated by the rise in the x standard deviation. It can be seen that as a result of the laser-based relative position information, there is a sharp decrease in the position standard deviations of robot #2 (marked by arrows). As the motion of the robot is primarily in the x direction when the corrections are provided, the resulting decrease in the x standard deviation is noticeable compared to those in y and ϕ.

Figure 5 shows a partially updated terrain map that was developed by two robots, Augustus and Theodosius, using the mapping procedure outlined in Section 4. Although this update is still performed offline for now, it shows the elevation profile across the area traversed by each robot, with prominent features within the robot's field of view during the motion segment being marked on the map.

Our multi-robot motion planning algorithm has been implemented in a 3D vehicle planner and control simulation environment. For typical multi-robot paths, collisions will occur if the paths are planned separately. Therefore velocity planning is necessary to resolve potential collisions. The velocity planning (D^* search in coordination diagram) on a typical $117 \times 95 \times 99$ grid took about 4 minutes. No consideration was given to reduce computation time in the software implementation. The velocity profiles for a typical example will give several solutions. For example, for a three-robot situation, one solution would be to insert three unit time delays for robot 2 at the beginning of its movement, a second solution is to insert four unit time delays for robot 3 at the beginning

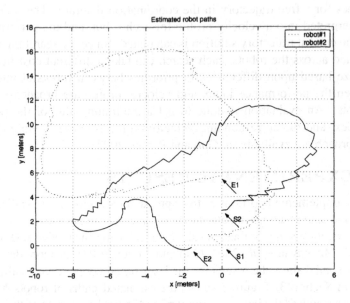

Figure 3. EKF estimated robot paths. The solid line denotes the estimated path of robot #2 and the dotted line that of robot #1. (S1,E1) and (S2,E2) denote the start and end positions for robots #1 and #2, respectively.

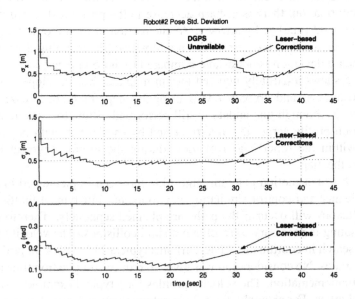

Figure 4. Laser-based cooperative localization, showing the standard deviation of the pose of robot #2. The external corrections offered by the laser-based localization scheme are marked by arrows.

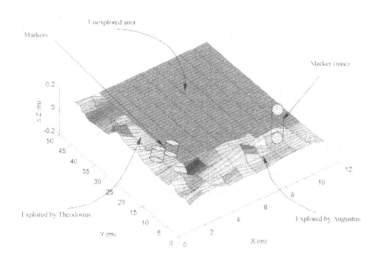

Figure 5. Partially updated terrain map.

of its movement, and a third solution is identical to the first. The differences in schedules are caused by assigning different priorities to robots. Since the first index is the smallest, the corresponding set of velocity profiles are chosen for each robot. It should be noted that although more complicated velocity profiles (with many stop-move schedules in the middle of the velocity profile) can be generated by the described algorithm, from a practical concern, based on the same or a comparable performance index value, it is preferred to have delays at the beginning of the velocity profiles, or to be consolidated, instead of requiring a lot of move-stop-move procedures during the robot movement. This can be achieved by smoothing zig-zag paths in the searching algorithm. Experimental work is underway to implement this algorithm on our group of ATRV-mini all-terrain mobile robots.

7. SUMMARY

In this paper, we have briefly outlined our approach toward using distributed heterogeneous sensing to achieve cooperative localization, mapping, and path planning in outdoor terrains using teams of mobile robots. We have sketched our algorithms toward achieving this goal and have given some initial results from implementation on our team of ATRV-mini robots. Our ultimate objective is to generate teams of mobile robots that can be placed in a previously unfamiliar outdoor environment, and use their distributed sensing capabilities to localize themselves, generate an approximate elevation map, and generate multi-robot paths that enable them to accomplish their intended objectives, such as perimeter security, reconnaissance and surveillance, and exploration.

References

Fregene, K., Madhavan, R., and Parker, L. E. (2002). Incremental multiagent robotic mapping of outdoor terrains. In *Proceedings of IEEE International Conference on Robotics and Automation*.

Guo, Y. and Parker, L. E. (2002). A distributed and optimal motion planning approach for multiple mobile robots. In *Proceedings of IEEE International Conference on Robotics and Automation*.

Matthies, L., Kanade, T., and Szelinski, R. (1989). Kalman Filter-based Algorithms for Estimating Depth from Image Sequences. *International Journal of Computer Vision*, 3:209–236.

Madhavan, R., Fregene, K., and Parker, L. E. (2002). Distributed heterogenous outdoor multi-robot localization. In *Proceedings of IEEE International Conference on Robotics and Automation*.

Roumeliotis, S. and Bekey, G. (2000). Distributed Multi-Robot Localization. In *Distributed Autonomous Robotic Systems - Chapter 6: Localization*, pages 179–188. Springer-Verlag.

MISSION-RELEVANT COLLABORATIVE OBSERVATION AND LOCALIZATION

Ashley W. Stroupe
Carnegie Mellon University, Pittsburgh, Pennsylvania, USA
ashley@ri.cmu.edu

Tucker Balch
Georgia Institute of Technology, Atlanta, Georgia, USA
tucker@cc.gatech.edu

Abstract: We propose an approach for the integration of collaborative observation and localization within an overall multi-robot mission. The proposed approach will allow individual robots within a team to evaluate and choose a task based on the needs of the mission and the needs, capabilities, and task choices of teammates. Tasks include mission-related tasks as well as task for self-localization, teammate-localization, and collaborative mapping. To date, work has been done in collaborative observation and mapping and robot localization.

Keywords: Multi-robot, collaborative tasks

1. INTRODUCTION

Near-term goals for multi-robot systems include applications in complex environments with large robot teams, such as planetary science, construction, surveillance/reconnaissance, and search and rescue. Successful completion of such complex tasks with limited resources and time will require taking full advantage of multi-robot systems' potential to improve efficiency, capability, and reliability, as well as addressing complex dynamic environments.

A.C. Schultz and L.E. Parker (eds.), Multi-Robot Systems: From Swarms to Intelligent Automata, 31-40.
© *2002 Kluwer Academic Publishers. Printed in the Netherlands.*

Suppose a robot team could create and maintain a consistent and accurate understanding of the environment while performing a high-level mission. We hypothesize that such a capability can substantially improve the system's effectiveness. Accuracy is required to perform individual tasks precisely and efficiently; consistency is required for effective cooperation among agents in tasks that cannot be accomplished individually. To achieve a consistent and accurate understanding of the environment, robots must make and combine multiple observations of objects and teammates and share results.

Integration of this collaborative mapping and localization with a complex mission increases the complexity of the task selection process. More tasks, capabilities and joint effects of teammates, and priorities of different tasks (which change with time and state) must be considered.

The proposed approach will deal with task selection, mapping, and localization, each in a multi-robot context. To date, work has been done in cooperative observation of static and dynamic targets and localization using *TeamBots* architecture and Minnow robots (details available at http://www.cmu.edu/~multirobotlab). Future work will include the design of explicitly collaborative localization and observation and design and implementation of a behavioral framework for incorporating localization and mapping into a multi-robot mission.

2.   RELATED WORK

Localization maps sensor data to robot pose using a representation of the environment. Approaches range from simple geometry to probabilistic modeling of pose estimates. Triangulation (Goel, *et al.*, 1999) uses range and bearing to landmarks to get the pose most consistent with sensing (and a motion model). More complex geometric approaches process data to create "images," Closest match and interpolation with stored templates provides pose (Lu and Milios, 1997). Most current approaches represent pose by a probability distribution or density. Occupancy grids assign discrete cells a probability they are occupied and are used like templates (Cai, *et al.*, 1996). Kalman-Bucy filters and Bayes' Rule filters use Gaussian distributions, allowing simple updates (Roumeliotis and Bekey, 2000). Markov and Monte Carlo Localization (MCL) approaches allow any type of distribution at the cost of complexity and/or resolution (Fox, *et al.*, 2000). Many approaches apply to small multi-robot systems. Using triangulation, teammates take turns exploring and serving as stationary landmarks (leap-frog) (Grabowski, *et al.*, 2000). In probabilistic approaches, probabilistic estimates of observed teammates poses are shared and used in updates (Roumeliotis and Bekey, 2000; Fox, *et al.*, 2000).

Mapping provides an environment representation for localization and task selection. This representation can be location of objects or landmarks or terrain elevations. Approaches are similar to localization, and mapping and localization are often done simultaneously (Borthwick, *et al.*, 1994; Dissanayake, *et al.*, 2000). Simple approaches map relative position of objects to world position using robot pose, typically in a discrete world. More complex approaches generate occupancy (or traversability) grids (Cai, *et al.*, 1996) or a probabilistic position estimate of objects using Bayes (Kalman-Bucy) (Borthwick, *et al.*, 1994) updates or Markov/MCL (Dissanayake, *et al.*, 2000). Probabilistic approaches are also used to track dynamic objects. Most multi-robot approaches reduce the problem into single-robot problems by dividing the area (Dias and Stentz, 2000) by or using leap-frog approaches (Fox *et al.*, 2000). Multiple robots can also jointly update a single grid cell occupancy map (Cai, *et al.*, 1996). Some work has examined how to deploy a team to maximize coverage without considering overlapping observations.

Members of a team must allocate mission tasks among agents. The simplest approach, applicable when a single task is required, is division of the task space (Dias and Stentz, 2000). Some approaches are dynamic, changing results as new information is discovered and faults occur. Task selection is typically based on the value of different tasks to a robot agent. The evaluation of value can be done in a reactive or behavior-based fashion, using predefined or learned functions (Darker, 1998), negotiation (Dias and Stentz, 2000), or traditional planning (Brumitt and Stentz, 1996).

The primary limitations of these approaches are lack of integration and inability to fully utilize the multi-robot system. Implementations of task selection algorithms focus on few choices, typically a single high-level goal, and do not incorporate pose or map uncertainty in the decision process. Systems make no explicit movement to obtain good team observations, and knowledge of teammate decision processes is not fully addressed.

3. APPROACH

The approach summarized here is detailed in (Stroupe, Sikorski, and Balch, 2001).

3.1 Task Selection

Robot tasks are specified by the mission directly (such as scientific analysis) or indirectly (such as localization and a map). The apparent most valuable task is the task selected. Task value depends on current priorities of different tasks, current task status, current state and pose estimate (including uncertainty), and resource availability. For example, mapping has higher value when map certainty is low, when other mission-related tasks have low

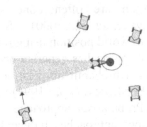

Figure 1.   Value functions. As pose uncertainty increases, localization's value increases and tasks requiring precision

Figure 2. Team observing dynamic target with uncertainty. As new data arrives, robots can refine their positions.

priority or value (due to low map quality), or when mapping sensors are more accurate. Value may also include an estimate of cost to complete the task. Unlike previous approaches, value may also be affected by teammate availability to participate in the task and the current needs of teammates. For example, localizing a teammate with a high-priority task will have higher value than localizing one with a low-priority task, and some mission tasks may have low value when other necessary resources are unavailable.

Given task selection partially depends on teammate choices, agreement must be reached among participating agents. This may lead to an iterative process of evaluating value. For efficiently selecting a task, the number of agents considered in evaluation and included in agreement may be reduced by using agent subsets defined by proximity, availability, or a priori groups.

Values for behaviors are defined by functions; the highest-valued function at any time selects the most appropriate behavior. Functions may be partially generic and partially specific to task and robot, but may also change as mission aspects become more or less important and the state changes. An example evaluation is in Figure 1. Mission tasks requiring accuracy become less valuable as pose uncertainty increases; at some uncertainty level, explicitly improving pose becomes the highest valued task and is selected.

3.2   Collaborative Mapping and Tracking

Collaborative mapping and observation attempts to take full advantage of the multi-robot team. Improved accuracy can be achieved by taking multiple observations on objects. Using individual sensor models, the value of observing at a point can be determined using the expected improvement from the observation and the cost to achieve that observation. By looking at sets of robot-position assignments, team deployment can be optimized for improving estimate quality. This can be done for static or dynamic targets.

Analytical optimization can be done for simple value functions over small numbers of robots and objects. For large teams and complex functions, acceptable solutions may be found using techniques such as hill-climbing, simulated annealing, spatial discretization, coarse-to-fine, and search space reduction based on proximity to position and targets. Coarse discrete solutions provide starting points for such approximate techniques.

For static targets, value may be evaluated as described. Efficiency is achieved through search space reduction and, if necessary, dividing larger teams into smaller teams. To leverage multiple observations by a single robot, selecting multiple points along or near a provided trajectory (defined by mission tasks, for example) reduces search.

For dynamic objects, when single robots cannot be assigned full-time tracking duty, iterative solutions become more important. The first solution estimate will depend on a preliminary position and velocity estimate, providing a range of area that may be traversed by the object. Initially, robots will take up vantage points around this area. As more information is collected, uncertainty is reduced and the area collapses, allowing agents to obtain more useful positions (Figure 2). The primary consideration is time: observation points lose value if they cannot be achieved before the target arrives. Also, value of observation points may need to include many potential target points.

3.3   Collaborative Localization

Collaborative localization includes both cooperative localization and collaboration (explicitly moving to aid teammates). When teammates encounter each other, each makes a relative and absolute position estimate of the other. The absolute estimate can be used by the other to update pose.

Should a teammate call for help, one or more robots must answer the call and attempt to localize the lost robot. The value of localizing the teammate depends on the value of the teammate's current task, the estimated value to other robots of localizing the teammate, sensor models, and travel distance and time required to sufficiently localize the teammate. Robots with high value on helping decide which robot(s) will proceed via negotiation. This decision depends on how uncertain the lost robot is and how far away the robot is from landmarks. Larger uncertainties require more measurements to correct and more robots to cover the possible area in which the teammate is. Also, if rescuers must drive far from landmarks, their uncertainty will grow making observations less valuable and travel more risky.

For reliability, localization must include ability to explicitly self-localize in case assistance is unavailable or very costly. The robot must maximize its chance of localizing by maximizing its chance of locating landmarks. The probability of locating landmarks depends on uncertainty in position and heading, chosen direction of travel, and distribution of landmarks within the area potentially traversed by choosing that direction. The resulting value of explicit self-localization depends directly on this probability.

4.   PROGRESS

4.1   Cooperative Observation

We have implemented and tested several aspects of cooperative observation, including the combining simultaneous observations of static and single dynamic objects from multiple robots (Stroupe, Martin, and Balch, 2001) and optimized selection of observation points of multiple static targets (Stroupe, Sikorski, and Balch, 2001). This approach combines observations with Bayes' Rule, using a Gaussian distribution to represent position and simple mathematical updates. By combining observations, the resulting quality of position estimates is improved.

4.1.1   Combining Observations

The approach first transforms local parameters directly obtained from sensor data into global parameters, using geometrical relations and a rotation transformation, so that observations may be combined using a simple matrix approach. The result is a covariance matrix and mean of the combined distribution (Smith and Cheeseman, 1986):

$$C' = C_1 - C_1 [C_1 + C_2]^{-1} C_1$$
$$\hat{X}' = \hat{X}_1 + C_1 [C_1 + C_2]^{-1} (\hat{X}_2 - \hat{X}_1)$$

This approach assumes that positional uncertainty is absent or previously incorporated. It also assumes that measurement distributions are Gaussian and that observations are simultaneous. These assumptions make processing very efficient and real-time even for large teams.

4.1.2 Distributed Sensing Experiments

Several distributed sensing experiments were conducted: locating static targets, tracking dynamic objects, and increasing effective field of view. In the first, an object is placed at a series of points and observed by three robots. Position estimates are combined together and in pairs. Resulting estimates provide more accurate results, with fewer outliers and lower maximum error (Figure 3). In the second, a robot is to push a box to a goal position; the box location is unseen and unknown. The robot receives a position estimate from a teammate (that cannot reach the object), allowing it to proceed to the object without searching and to manipulate it with a more accurate combined estimate. In the third experiment, three robots visually track a target by using the combined position estimate from all three observations. When one robot is blindfolded, it can still track the target using teammate observations. This approach was used in RoboCup 2000, allowing increased field coverage.

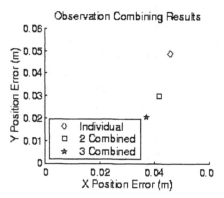 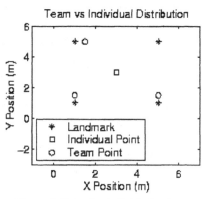

Figure 3. Mean error in estimates from single and multiple observations combined. Each additional observation reduces error.

Figure 4. Selecting individual points puts all robots at the same place. Selecting for a team distributes robots to improve result.

4.1.3 Optimized Mapping Simulations

For optimized mapping, a team of robots is deployed to best improve a landmark map by optimizing expected reward and cost. Expected reward uses sensor models, landmark positions, and the method of combining observations to predict the resulting from all observations. Expected cost uses driving time and distance. Robot pose uncertainty is ignored, as it is hard to predict. Simulation results show resulting position estimates improve on multi-robot estimates with individually chosen observation points (Figure 4).

Simulations verify results are highly dependent on value function and reward/cost relative weight; values must be tuned to team and application.

4.2   Localization

Driven by the need for a localization approach requiring minimal processing and applicable with minimal (and noisy) landmark information and availability, we have developed an implemented a constraint-based approach to landmark localization (Smith, *et al.*, 1986). Localization has been implemented for individual robots but can easily include teammates as landmarks.

4.2.1   Localization Updates

Different types of measurements on landmarks lead to geometric constraints on robot position and heading. A range provides a circular constraint around a point landmark while a bearing to a point landmark provides a single heading value and a linear constraint on position. These constraints are easily computed algebraically. Each independent estimate can be combined, along with previous estimates and motion models, to efficiently determine a new pose estimate. In this way, all available information is used without requiring the same quantity or quality of information from all landmarks or measurements. The current implementation does not provide an estimate of result quality. Additionally, the implementation does not use teammates as landmarks (robots cannot yet accurately sense teammate pose), though this could be easily incorporated.

4.2.2   Localization Experiments

The localization approach is tested by driving a robot along a circuit, defined by waypoints, among landmarks. As the robot reaches each waypoint, it records its pose as reported by on-board odometry and by localization; ground truth is also measured. Experiments were conducted in which number and type (for position estimation) of landmarks was varied.

Localization eliminates the typical odometry drift, even with very few landmarks providing only range for position estimation (Figure 5). As more information is added (such as bearing and more landmarks), performance further improves. Position and heading errors are non-zero due to noisy sensing, but are consistent through time and do not grow like odometry.

The approach was applied in RoboCup middle-size soccer, 2001. This environment provides only four bearing-only landmarks, and no more than two are visible together. Qualitative results demonstrate drift is eliminated;

robots returning to home position did not drift toward the field edge or lose ability to orient down-field. In practice, errors remained on the order of 0.5 m and 10 degrees after 10 minutes of play, much reduced from the 2 m and 45 degrees (and beyond) errors in odometry in the same time.

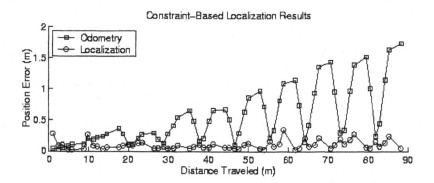

Figure 5. Localization eliminates odometry drift even with few, low-quality landmarks.

5.   CONCLUSION

We propose an approach to integrate collaborative localization and observation with an overall, complex mission. Robots select an applicable task based on the current value of available tasks, including mapping and localization of self and of teammates. Mapping is done by optimizing multiple simultaneous observation points for available teammates, providing a team-optimized result. Dynamic object observation optimizes multiple, time-distributed observation points. Collaborative localization utilizes fortuitous encounters and explicit movement to localize self or teammates.

Preliminary results in combining multiple observations, collaborative observation point selection, and localization indicate these separate aspects are quite promising. Further work will be done in developing an integrated collaborative multi-robot system, including developing the framework for task selection and value functions for some mission tasks, mapping, and localization. This also includes reducing the search space for team deployment in observation tasks to provide acceptable results efficiently. The system will be implemented and tested in simulation and on robot teams.

References

Borthwick, S. and Durrant-Whyte, H. (1994). Simultaneous Localisation and Map Building for Autonomous Guided Vehicles. In *Proceedings of IROS*.

Brumitt, B. and Stentz, A. (1996). Dynamic Mission Planning for Multiple Mobile Robots. In *Proceedings of ICRA*.

Cai, A., Fukuda, T., Arai, F., and Ishihara, H. (1996). Cooperative Path Planning and Navigation Based on Distributed Sensing. In *Proceedings of ICRA*.

Dias, M. B. and Stentz, A. (2000). A Free Market Approach to Distributed Control of a Multirobot Coordination. In *Proceedings of 6th Int. Conference on Intelligent Autonomous Systems*.

Dissanayake, G., Durant-Whyte, H., and Bailey, T. (2000). A Computationally Efficient Solution to the Simultaneous Localisation and Map Building (SLAM) Problem. In *Proceedings of ICRA*.

Fox, D., Burgard, W., Kruppa, H., and Thrun, S. (2000). A Probabilistic Approach to Collaborative Multi-Robot Localization. *Autonomous Robots Special Issue on Heterogeneous Multi-Robot Systems*, 8(3).

Goel, P., Roumeliotis, S. I., and Sukhatme, G. S. (1999). Robust Localization Using Relative and Absolute Position Estimates. In *Proceedings of IROS*.

Grabowski, R., Navarro-Serment, L.E., Pareidis, C. J. J., and Khosla, P. K.. Heterogeneous Teams of Modular Robots for Mapping and Exploration. *Autonomous Robots Special Issue on Heterogeneous Multi-Robot Systems*, 8(3).

Lu, F. and Milios, E. (1997). Robot Pose Estimation in Unknown Environments by Matching 2D Range Scans. *Journal of Intelligent Robotic Systems*, 18.

Parker, L. (1998). ALLIANCE: an architecture for fault tolerant multirobot cooperation. *IEEE Transactions on Robotics and Automation*, 14(2).

Roumeliotis, S. I. and Bekey, G. A. (2000). Synergetic Localization for Groups of Mobile Robots. In *Proceedings of IEEE Conference on Decision and Control, 2000*.

Smith, R. C. and Cheeseman, P. (1986). On the Representation and Estimation of Spatial Uncertainty. *International Journal of Robotics Research*, 5(4). Stroupe, A. W., Martin, M. C., and Balch, T. Distributed Sensor Fusion for Object Position Estimation by Multi-Robot Systems. In *Proceedings of ICRA*.

Stroupe, A. W., Martin, M. C., and Balch, T. Distributed Sensor Fusion for Object Position Estimation by Multi-Robot Systems. In *Proceedings of ICRA*.

Stroupe, A., Sikorski, K., and Balch, T. (2001). *Mission-Driven Collaborative Observation and Localization*. Thesis Proposal, Carnegie Mellon University.

DEPLOYMENT AND LOCALIZATION FOR MOBILE ROBOT TEAMS

Andrew Howard and Maja J Matarić

Robotics Research Laboratory, Computer Science Department
University of Southern California

ahoward@usc.edu, mataric@usc.edu

Abstract   This paper briefly sketches a pair of algorithms for deploying and localizing large mobile robot teams. For deployment, we have developed a potential-field-based approach that ensures that a compact initial configuration of robots will spread out such that the area 'covered' by the team is maximized. For localization, we have developed an approach that makes use of the robots themselves as landmarks. Through a combination of maximum likelihood estimation and numerical optimization, we can, for each robot, estimate the relative range, bearing and orientation of every other robot in the team. This paper sketches the basic formalism behind these algorithms and presents some experimental results.

Keywords: Robot teams, deployment, localization.

1.   INTRODUCTION

We are interested in the use of mobile robot teams in applications ranging from urban combat, to search-and-rescue and environment monitoring. Consider, for example, a search-and-rescue scenario in which a robot team must deploy into a damaged structure, locate survivors, and guide rescuers to those survivors. This paper addresses two of the key problems that must be solved in order to conduct such missions: deployment and localization.

The first problem we consider is that of localization; i.e. how does each robot determine its pose with respect to every other robot in the team. In the scenario we have described, localization information cannot be obtained using GPS or landmark-based techniques; GPS is generally unavailable or unreliable in urban environments due to multi-path effects, while landmark-based techniques require prior models of the environment that are either unavailable, incomplete or inaccurate. For these reasons, we have developed an approach to localization that relies on using the robots *themselves* as landmarks. Robots are equipped with sensors that allow them to measure the relative pose of nearby

A.C. Schultz and L.E. Parker (eds.), Multi-Robot Systems: From Swarms to Intelligent Automata, 41-51.

robots, and sensors that allow them to measure changes in their own pose. Given the measurements from these sensors, a combination of maximum likelihood estimation and numerical optimization is used to determine the most probable pose for each robot. With this approach, one can obtain good localization information in almost any environment, including those that are undergoing dynamic structural changes. Our only requirement is that the robots are able to maintain at least intermittent line-of-sight contact with one another.

The second problem we consider is that of deployment; i.e., how does one control the motion of robots such that the team as a whole maintains or optimizes some desired set of properties. Two properties are of particular interest: area coverage and line-of-sight connectivity. The former property is clearly useful for scenarios such as the one described above, and the latter property is *required* if the localization technique described in this paper is to be effective. As with localization, the deployment algorithm is constrained in that it cannot make use of prior models of the environment. Therefore, we have developed an approach to deployment that relies entirely on sensed rather than stored data. This approach makes use of virtual potential fields; i.e., robots are treated as virtual particles that are subject to virtual forces; these forces repel the robots from one other and from obstacles. The forces are such that an initial, compact configuration of robots will eventually spread out to cover a much larger area of the environment. Both area coverage and line-of-sight connectivity are emergent properties of this algorithm.

It should be noted that the localization and deployment algorithms described in this paper are entirely distributed. We regard this as necessary property for any algorithm that must scale to teams with thousands or tens-of-thousands of robots.

The remainder of this paper is divided into two main sections, the first of which treats the problem of localization, and the second of which treats the problem of deployment.

2.   LOCALIZATION

Our approach to team localization relies on two basic assumptions. First, we assume that each robot is equipped with a proprioceptive *motion detector* such that it can measure changes in its own pose (subject to some degree of uncertainty). Suitable motion detectors can be constructed using either odometry or inertial measurement units. Second, we assume that each robot is equipped with a *robot detector* such that it can measure the relative pose and identity of nearby robots. Suitable sensors can be constructed using either vision or scanning laser range-finders. We further assume that the identity of robots is always measured correctly (which eliminates what would otherwise be a com-

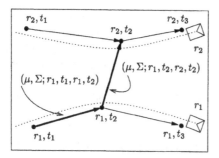

Figure 1. Illustration of the basic formalism using a directed graph. The figure shows two robots, r_1 and r_2, traveling from left to right; at time t_2, robot r_1 observes robot r_2. The vertices in the graph represent pose estimates; the edges represent observations. Two observations are highlighted: a motion observation and a robot observation, both made by robot r_1.

binatoric labeling problem) but that there is some uncertainty in the relative pose measurements.

Given these assumptions, the team localization problem can be solved using maximum likelihood estimation. The basic method is as follows. First, we construct a set of estimates $H = \{h\}$ in which each element h represents a pose estimate for a particular robot at a particular time. These pose estimates are defined with respect to some *arbitrary* global coordinate system. Second, we construct a set of observations $O = \{o\}$ in which each element o represents a relative pose measurement made by either a motion or robot detector. For motion detectors, each observation o represents the measured change in pose of a single robot; for robot detectors, each observation o represents the measured pose of one robot relative to another. Finally, we use numerical optimization to determine the set of estimates H that is most likely to give rise to the set of observations O.

One can visualize the approach in terms of a directed graph, as shown in Figure 1. We associate each estimate h with a vertex in the graph, and each observation o with an edge. Each vertex can have both outgoing edges, corresponding to observations in which this vertex was the *observer*, and incoming edges, corresponding to observations in which this vertex was the *observee*. Each edge also has a natural 'length', corresponding to the measured relative pose of the two vertices. Numerical optimization is used to determine the configuration of vertices that mimimizes the amount of 'stretching' in the edges; i.e. to minimize the difference between the *estimated* and *measured* relative pose for each pair of vertices.

Note that, in general, we do not expect robots to use the set of pose estimates H directly; these estimates are defined with respect to an arbitrary coordinate system whose relationship with the external environment is undefined. Instead, each robots uses these estimates to compute the pose of every other robot *rel-*

ative to itself, and uses this ego-centric viewpoint to coordinate activity. We note, however, that some subset of the team may choose to remain stationary, thereby 'anchoring' the global coordinate system in the real world. In this case, the pose estimates in H may be used as global coordinates in the standard fashion.

The detailed mathematical formalism underlying this approach can be found in (Howard et al., 2001a).

2.1   From Centralized to Distributed Implementations

The formalism described above can be readily implemented using a *centralized* algorithm. Observations from all robots are reported to a central base-station, where an efficient numerical optimization algorithm is applied to the data (we typically use a conjugate gradient algorithm). Estimated poses are communicated back to the individual robots. While this algorithm may be suitable for applications in which distances are short and robots are few, it does not scale well to large networks. This algorithm has both computation and communication bottlenecks.

The formalism can, however, be implemented in an entirely *distributed* fashion. Consider once again the directed-graph visualization in Figure 1. We can divide this graph into a set of fragments, each of which encompasses the poses estimates for a single robot. Each robot is then made responsible for its 'own' fragment, and conducts a separate numerical optimization. The robots must periodically broadcast their set of pose estimates (to ensure global consistency) and must explicitly communicate robot observations (since the corresponding edges are effectively 'shared' by two fragments).

At present, this distributed algorithm is largely speculative (it has not been implemented). Our hope is that the algorithm will have both constant-time and constant-bandwidth properties (using short-range, local communications), and will therefore scale to networks of any size.

2.2   Results

Figure 2 shows the results for an experiment conducted using a team of 7 robots. Each robot is comprised of a Pioneer 2DX mobile robot, a SICK LMS 200 scanning laser range-finder and a retro-reflective bar-code. Motion observations are provided by the robot's on-board odometry; robot observations are provided by the laser range-finder/bar-code combination. For this experiment, 6 of the 7 robots were positioned at fixed locations in the corridors of a building, as shown in Figure 2(a); the remaining robot was then 'joy-sticked' around the circuit, and was thus 'seen' by each of the stationary robots in turn.

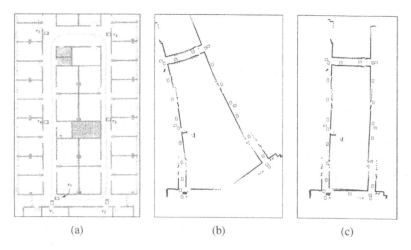

(a)	(b)	(c)

Figure 2. An experiment with real robots. (a) Experimental set-up: six stationary robots (r_1 to r_6) are placed at strategic locations; the seventh mobile robot (r_0) executes a circuit. (b) Combined laser scans at $t = 200$ sec, after the mobile robot has been seen by all six stationary robots exactly once. Note that this is *not* a stored map: this is live laser data. Pose estimates and observations are also shown, denoted by rectangles and lines respectively. (c) Combined laser scans at $t = 220$ sec, after the mobile robot has been seen by the first stationary robot r_1 for a second time, thus closing the loop.

The stationary robots were positioned outside each other's sensor range, and hence there are no observations that relate these robots directly.

Figure 2b shows the combined laser scans at $t = 200$ sec, after the mobile robot has been seen by all six stationary robots exactly once. Note that this is *not* a stored map: this is live laser data. The cummulative error in the mobile robots's odometry has clearly manifested itself as a 'bend' in this plot. In contrast, Figure 2(c) shows the combined laser scans at $t = 220$ sec, after the mobile robot has been seen by the first stationary robot r_1 for a second time, thus 'closing the loop'. The cummulative error has been erased.

These result were generated using a centralized algorithm. More plots, with associated quantative analysis, can be found in (Howard et al., 2001b) and (Howard et al., 2001a).

2.3 Related Work

Localization is an extremely well studied area in mobile robotics. The vast majority of this research has concentrated on two problems: localizing a single robot using an a priori map of the environment (Leonard and Durrant-Whyte, 1991; Simmons and Koenig, 1995; Fox et al., 1999), or localizing a single robot whilst simultaneously building a map (Thrun et al., 1998; Lu and Milios, 1997; Yamauchi et al., 1998; Duckett et al., 2000; Golfarelli et al.,

1998; Dissanayake et al., 2001). Recently, some authors have also considered the related problem of map building with multiple robots (Thrun, 2001). All of these authors make use of statistical or probabilistic techniques of one sort or another; the common tools of choice are Kalman filters, maximum likelihood estimation, expectation maximization, and Markovian techniques (using grid or sample-based representations for probability distributions). The team localization problem described in this paper bears many similarities to the simultaneous localization and map building problem, and is amenable to similar mathematical treatments. Our mathematical formalism is very similar, for example, to that described in (Lu and Milios, 1997).

Among those who have considered the specific problem of team localization are (Roumeliotis and Bekey, 2000) and (Fox et al., 2000). Roumeliotis and Bekey present an approach to multi-robot localization in which sensor data from a heterogeneous collection of robots is combined through a single Kalman filter to estimate the pose of each robot in the team. This method relies entirely on external landmarks; no attempt is made to sense other robots or to use this information to constrain the pose estimates. In contrast, Fox *et al.* describe an approach to multi-robot localization in which each robot maintains a probability distribution describing its own pose (based on odometry and environment sensing), but is able to refine this distribution through the observation of other robots. It is not clear, however, that this technique can be applied to problems in which *only* the robots are used as landmarks.

Finally, a number of authors (Kurazume and Hirose, 2000; Rekleitis et al., 1997) have considered the problem of team localization from a somewhat different perspective. These authors describe *cooperative* approaches to localization, in which team members actively coordinate their activities in order to reduce cumulative odometric errors. The basic method is to keep one subset of the robots stationary, while the other robots are in motion; the stationary robots observe the robots in motion (or vice-versa), thereby providing more accurate pose estimates than can be obtained using odometry alone. While our approach does not require such explicit cooperation on the part of robots, the accuracy of localization can certainly be improved by the adoption of such strategies.

3.   DEPLOYMENT

We have developed a potential-field-based approach to deployment. In this approach, robots are treated as virtual particles that are subject to virtual forces; these forces repel the robots from each other and from obstacles. In addition, robots are also subject to a dissipative viscous force; this force is used to ensure that the team will eventually reach a state of static equilibrium; i.e. all robots will ultimately come to a complete stop. The viscous force does not, however, prevent the team from reacting to *changes* in the environment; if something is

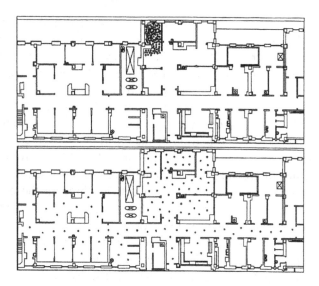

Figure 3.   A proto-typical deployment experiment for a 100-robot network. Top: initial network configuration. (b) Bottom: final configuration after 300 seconds.

moved, the team will automatically reconfigure itself for the modified environment before once again returning to a state of static equilibrium. Thus, robots move only when it is necessary to do so, which saves a great deal energy.

Our only assumption is that each robot is equipped with a sensor, such as a scanning laser range-finder or omni-camera, that allows it to determine the range and bearing of both nearby robots and obstacles. Using this information, the robot can determine the virtual forces acting it, and convert this information into a control vector to be sent its motors. No other information is required.

It should be noted that this approach *does not* require global localization or communication between robots. While both capabilities may in fact be present in the network, we would prefer to construct a minimalist system that makes few assumptions about the environment and relies on few capabilities on the part of the robots. This is partly for the sake of producing an approach that is robust, and partly for the sake of creating a base-line system against which more complex and more capable techniques can be compared.

The detailed mathemathical formalism underlying this approach can be found in (Howard et al., 2002).

3.1   Results

Figure 3 shows a typical deployment conducted in using the Stage multi-agent simulator (Gerkey et al., 2001). For this experiment, the simulated team consisted of 100 robots, each of which is equipped with a scanning laser range finder, a retro-reflective beacon and an omni-directional mobile robot base. The laser has a 360 degree field-of-view and can determine the range and bearing of objects out to a range of 4m. The laser can also distinguish between robots (which carry a retro-reflective beacon) and obstacles (which do not). The network was placed in a large, complex, simulated hospital environment.

Figures 3 shows the initial and final network configurations for this experiment. From their starting configuration (packed into a single room) the robots spread out to cover a sizeable portion of the environment; the coverage area in the final configuration is in excess of 500 m^2, a 10-fold improvement over the initial coverage of around 50 m^2. The deployment also such that the network maintains complete line-of-sight connectivity throught the duration of the experiment. That is, every robot lies within the sensory field of at least one other robot, and any pair of robots in the network can be connected via a series of such line-of-sight relationships. Full connectivity is an emergent property of the deployment, and it is unclear at this time how reliably this property emerges. While the potential field algorithm is such that each robot must remain within the sensory range of at least one other robot (since a robot cannot be repelled by a robot it cannot see), this is not sufficient to guarantee full connectivity. One could, for example, imagine a situation in which the teams breaks into multiple disconnected islands. This topic remains the subject of further investigation.

Futher information can be found in (Howard et al., 2002).

3.2   Related Work

Potential field techniques for robotic applications were first described by Khatib (Khatib, 1986) and have since been widely used in the mobile robotics community for tasks such as local navigation and obstacle avoidance. The related concept of 'motor schemas', which utilizes the super-position of spatial vector fields to generate behavior was introduced by Arkin (Arkin, 1989). Both techniques have since been applied to the problem of formation control for groups of mobile robots (Scheider et al., 2000; Balch and Hybinette, 2000). The formation problem is similar, in some respects, to the deployment problem described in this paper, in that the robots will attempt to maintain a formation based on local sensing and computation. A key difference, however, is that there is no requirement that the formation reach a state of static equilibrium.

The deployment problem also is also similar, in some respects, to the multi-robot exploration and mapping problem. Here, the aim is to build a global map of the environment by sequentially visiting each location with one or more robots. This problem has been considered by a number of authors (Dedeoglu and Sukhatme, 2000; Simmons et al., 2000; Burgard et al., 2000) who use a variety of techniques ranging from topological matching (Dedeoglu and

Sukhatme, 2000) to fuzzy inference (López-Sánchez et al., 1998) and particle filters (Thrun et al., 2001).

Finally, we note that the concept of *coverage* as a paradigm for evaluating many-robot systems was introduced by Gage (Gage, 1992). Gage defines three basic types of coverage: blanket coverage, where the objective is to achieve a static arrangement of robots that maximizes the total detection area; barrier coverage, where the objective is to minimize the probability of undetected penetration through the barrier; and sweep coverage, which is more-or-less equivalent to a moving barrier. According to this taxonomy, the deployment problem described in this section is a blanket coverage problem.

4. FUTURE DIRECTIONS

To date, we have treated the problems of localization and deployment quite separately. There is much scope, however, for combining these two problems; for example, one can design deployment strategies that explicitly maximize the accuracy of localization by reasoning about uncertainty.

We are also considering two additional problems: recovery and self-repair. Ideally, once the team has performed its task, we would like to be able to recover, or 'un-deploy' the team. Interestingly, this is a more difficult task than deploying the team in the first place (one can see that this is likely to be the case using a basic entropy argument). Similarly, since the team is intended for hostile environments in which robots may be destroyed (either by accident or malicious activity), we would like the team to be able to repair itself. This can be done by reposition robots to patch the 'holes' formed by the destroyed robots.

References

Arkin, R. C. (1989). Motor schema based mobile robot navigation. *International Journal of Robotics Research*, 8(4):92–112.

Balch, T. and Hybinette, M. (2000). Behavior-based coordination of large-scale robot formations. In *Proceedings of the Fourth International Conference on Multiagent Systems (ICMAS '00)*, pages 363–364, Boston, MA, USA.

Burgard, W., Moors, M., Fox, D., Simmons, R., and Thrun, S. (2000). Collaborative multi-robot exploration. In *Proc. of IEEE International Conferenceon Robotics and Automation (ICRA)*, volume 1, pages 476–81.

Dedeoglu, G. and Sukhatme, G. S. (2000). Landmark-based matching algorithms for cooperative mapping by autonomous robots. In Parker, L. E., Bekey, G. W., and Barhen, J., editors, *Distributed Autonomous Robotics Systems*, volume 4, pages 251–260. Springer.

Dissanayake, M. W. M. G., Newman, P., Clark, S., Durrant-Whyte, H. F., and Csorba, M. (2001). A solution to the simultaneous localization and map building (slam) problem. *IEEE Transactions on Robotics and Automation*, 17(3):229–241.

Duckett, T., Marsland, S., and Shapiro, J. (2000). Learning globally consistent maps by relaxation. In *Proceedings of the IEEE International Conference on Robotics and Automation*, volume 4, pages 3841–3846.

Fox, D., Burgard, W., Kruppa, H., and Thrun, S. (2000). A probabilistic approach to collaborative multi-robot localization. *Autonomous Robots, Special Issue on Heterogeneous Multi-Robot Systems*, 8(3):325–344.

Fox, D., Burgard, W., and Thrun, S. (1999). Markov localization for mobile robots in dynamic environments. *Journal of Artificial Intelligence Research*, 11:391–427.

Gage, D. W. (1992). Command control for many-robot systems. In *AUVS-92, the Nineteenth Annual AUVS Technical Symposium*, pages 22–24, Hunstville Alabama, USA. Reprinted in Unmanned Systems Magazine, Fall 1992, Volume 10, Number 4, pp 28-34.

Gerkey, B., Vaughan, R., and Howard, A. (2001). Player/Stage homepage. http://www-robotics.usc.edu/player/.

Golfarelli, M., Maio, D., and Rizzi, S. (1998). Elastic correction of dead reckoning errors in map building. In *Proceedings of the IEEE/RSJ International Conference on Intelligent Robots and Systems*, volume 2, pages 905–911.

Howard, A., Matarić, M. J., and Sukhatme, G. S. (2001a). Localization for mobile robot teams: A maximum likelihood approach. Technical Report IRIS-01-407, Institute for Robotics and Intelligent Systems Technical Report, University of Sourthern California.

Howard, A., Matarić, M. J., and Sukhatme, G. S. (2001b). Relaxation on a mesh: a formalism for generalized localization. In *Proceedings of the IEEE/RSJ International Conference on Intelligent Robots and Systems (IROS01)*, pages 1055–1060.

Howard, A., Matarić, M. J., and Sukhatme, G. S. (2002). Mobile sensor network deployment using potential fields: A distributed, scalable solution to the area coverage problem. In *Proceedings of the 6th International Conference on Distributed Autonomous Robotic Systems (DARS02)*, page to appear, Fukuoka, Japan.

Khatib, O. (1986). Real-time obstacle avoidance for manipulators and mobile robots. *International Journal of Robotics Research*, 5(1):90–98.

Kurazume, R. and Hirose, S. (2000). An experimental study of a cooperative positioning system. *Autonomous Robots*, 8(1):43–52.

Leonard, J. J. and Durrant-Whyte, H. F. (1991). Mobile robot localization by tracking geometric beacons. *IEEE Transactions on Robotics and Automation*, 7(3):376–382.

López-Sánchez, M., Esteva, F., de Mántaras, R. L., Sierra, C., and Amat, J. (1998). Map generation by cooperative low-cost robots in structured unknown environments. *Autonomous Robots*, 5(1):53–61.

Lu, F. and Milios, E. (1997). Globally consistent range scan alignment for environment mapping. *Autonomous Robots*, 4:333–349.

Rekleitis, I. M., Dudek, G., and Milios, E. (1997). Multi-robot exploration of an unknown environment: efficiently reducing the odometry error. In *Proc. of the International Joint Conference on Artificial Intelligence (IJCAI)*, volume 2, pages 1340–1345.

Roumeliotis, S. I. and Bekey, G. A. (2000). Collective localization: a distributed kalman filter approach. In *Proceedings of the IEEE International Conference on Robotics and Automation*, volume 3, pages 2958–2965.

Scheider, F. E., Wildermuth, D., and Wolf, H.-L. (2000). Motion coordination in formations of multiple mobile robots using a potential field approach. In Parker, L. E., Bekey, G. W., and Barhen, J., editors, *Distributed Autonomous Robotics Systems*, volume 4, pages 305–314. Springer.

Simmons, R., Apfelbaum, D., Burgard, W., Fox, D., Moors, M., Thrun, S., and Younes, H. (2000). Coordination for multi-robot exploration and mapping. In *Proc. of the Seventeenth National Conference on Artificial Intelligence (AAAI-2000)*, pages 852–858.

Simmons, R. and Koenig, S. (1995). Probabilistic navigation in partially observable environments. In *Proceedings of International Joint Conference on Artificial Intelligence*, volume 2, pages 1080–1087.

Thrun, S. (2001). A probabilistic online mapping algorithm for teams of mobile robots. *International Journal of Robotics Research*, 20(5):335–363.

Thrun, S., Fox, D., and Burgard, W. (1998). A probabilistic approach to concurrent mapping and localisation for mobile robots. *Machine Learning*, 31(5):29–55. Joint issue with Autonomous Robots.

Thrun, S., Fox, D., Burgard, W., and Dellaert, F. (2001). Robust monte carlo localization for mobile robots. *Artificial Intelligence Journal*, 128(1–2):99–141.

Yamauchi, B., Shultz, A., and Adams, W. (1998). Mobile robot exploration and map-building with continuous localization. In *Proceedings of the 1998 IEEE/RSJ International Conference on Robotics and Automation*, volume 4, pages 3175–3720.

MULTIPLE AUTONOMOUS ROBOTS FOR UXO CLEARANCE, THE BASIC UXO GATHERING SYSTEM (BUGS) PROJECT

Tuan N. Nguyen, Christopher O'Donnell, and Tuan B. Nguyen
Naval EOD Technology Division
2008 Stump Neck Road
Indian Head, MD 20640-5070
{nguyen, odonnell}@eodpoe2.navsea.navy.mil

Abstract: The Naval Explosive Ordnance Disposal Technology Division (NAVEODTECHDIV) has had an active program for several years for the development of technologies required to realize an autonomous system of small robots to clear areas of unexploded submunition. The focus, thus far, has been on the technology elements themselves, with an emphasis on the multiple robot operation approach and autonomous cooperative-behavior control system and processing. NAVEODTECHDIV is in full-scale testing of the system operation and performance.

Keywords: UXO, autonomy, robots, UGV, control

1.   INTRODUCTION

One of the most prevalent threats to Explosive Ordnance Disposal (EOD) personnel is scatter-able submunitions. These small explosive ordnance items are dispersed by the hundreds and even thousands, by various artillery and missile systems. They are often found on the surface and sometimes a number of these submunitions remain as unexploded ordnance (UXO) because they failed to function as intended. Current EOD disruption tools and techniques require the technician to remain in very close proximity to the UXO when performing Pick Up and Carry Away (PUCA) or Blow In Place (BIP). Standoff distance will remove the technician from possible harm if the

A.C. Schultz and L.E. Parker (eds.), Multi-Robot Systems: From Swarms to Intelligent Automata, 53-61.
© *2002 Kluwer Academic Publishers. Printed in the Netherlands.*

UXO initiates accidentally. These sensors include acoustic, anti-disturbance, vibration, infrared (active or passive), magnetic, photocell, pressure, seismic, and trip wires that can all be triggered by personnel or EOD robotic platforms working in close proximity to the threat. The BUGS will provide an alternative to current EOD PUCA or BIP techniques that require the EOD technician to perform. Furthermore, future conflicts have the potential to create a UXO problem that quickly surpasses current EOD manpower capabilities. The NAVEODTECHDIV is tasked to develop tools, techniques, and procedures to improve the capabilities of military EOD technicians from each of the four military services. In response to the submunition UXO threat, the Basic UXO Gathering System (BUGS) program was initiated based on a swarm of multiple small, inexpensive, autonomous robots that can provide an alternative to current EOD PUCA or BIP techniques that require the EOD technician to perform.

2.   THE BUGS PROGRAM

The objective of the BUGS program is to develop, test, evaluate, and demonstrate the use of multiple small robots in clearing land areas of small submunition UXO. Some key features of the notional BUGS concept include low cost per robot, small in size, and a high degree of autonomy. The low cost is important because a complete system may have fifty or more robots, so the acquisition cost of the system will be driven by individual robot cost. Also, due to the hazardous nature of the robot's task, it is expected that some robots will be destroyed as sensitive targets are disturbed. The occasional loss of an individual robot can be justified through its low cost. Small size is important because of logistical considerations. EOD typically operates in small squads, with modest transportation available, therefore the system needs to fit either into the back of a HMMWV or a small trailer to be pulled by the HMMWV. A high degree of robotic autonomy is required, because EOD operator(s) are expected to command a swarm of multiple robots, therefore individual robot cannot receive or demand the operator's full attention or time. The EOD operator should only become actively involved in rare situations when a problem arises.

3.   TECHNICAL PROGRAM

The BUGS program is focused on (1) system software control and development of maturate enabling technologies, (2) use of test bed hardware to explore different system implementations and (3) collecting information on costs, capabilities, and alternatives that feed to UXO Submunition Clearance Analysis of Alternatives. Two prototype systems are being developed, built, and tested under this program. The first system being developed by the Navy (figure 1) is for random search, focusing on the system simplicity and low cost. The in-house system focused on autonomous operation with a) no man in the loop b) no precise navigation c) no assisting asset and/or operator control unit (OCU). The second system being developed under contract, (Figure 2) focused on direct search and full area coverage. This system is more complex, but has more capabilities. The system is being built by General Dynamic Robotic System (GDRS-RST) focusing on a) Semi-autonomous – with user interaction capability b) Precise navigation (DGPS) - for full area coverage and c) Centralize control - for the multiple robots planning and control.

Figure 1. Naval EOD Technology Division system

Figure 2. GDRS-RST system

Full autonomy is the goal; limited man-in-the-loop operation is desired. Truly autonomous robots operating purposefully in unstructured environments currently do not exist. Both sets of robots have appropriate levels of sensing, autonomy, manipulation, and communications technologies to explore possibilities for BUGS.

The primary technological emphasis for the BUGS project is on its software development and hardware control. Keys technical challenge for the BUGS system configurations include - low-cost processing capability, robotic navigation, robotic communications, UXO target sensing, and long-endurance energy supplies, which influence the size and cost of these small, expendable robots. Within these two systems, a range of possibilities with varying costs and capabilities can be explored. These systems are being built, tested for system performance, and evaluated.

3.1 NAVEODTECHDIV System and Its Operational Approach (Random Search)

The multiple robot system is a tightly integrated system comprised of two major components: multiple robots and an electronic fence or GPS receiver containment system. Each robot's processing power is based on a single board computer, Motorola MC68332.

The random system does not have a-priori knowledge of knowing where the targets are located, but they know the designated area that they should work within. The robots will conduct a random search of the area since they do not have precision navigation capability. They will avoid obstacles and each other by using either short-range infrared, ultrasonic, and/or physical bumpers. Once the target is detected, the robot will perform (BIP) or (PUCA) by carrying the target to the fence boundary area or a general drop point area. Once the dropping task is done, the robot returns to its initial search mode, or wait to be called home. The random system behavior approach was developed based on the real world, where the system cannot perform precise navigation in rough terrains.

This robotic system takes advantage of simple decentralized rather then centralized, emergent rather than planned, and concurrent rather than sequential operational concepts. The decentralized system is based on having each robot perform the task autonomously and independently without depending on the operator or higher control system. The robot acts independently according to its ability in responding to its local view of the environment. The emergent behavior occurs through dynamic interaction of the robots and its terrain rather than globally planned patterns by the central OCU. The concurrence of the system is based on parallel operation rather than sequentially control from higher order systems.

3.2 GDRS System and Its Operation Approach (Direct and Pattern Search)

The system consists of a group of small intelligent vehicles coupled via a wireless local area network with an operator control station to help with the coordination of the multiple autonomous robots. The operator control unit, embodied in a laptop computer is used for operation planning, path planning, traffic management, and progress monitoring. The OCU initially downloads the mission plan to the robots, and then it tracks the robots locations, solves system conflicts, and prioritizes the multiple robots operation.

Each vehicle has a single board Pentium (Tillamook) computer running the real-time operating system QNX, as well as the following elements:

- Differential global positioning system (DGPS), with 20 cm nominal accuracy
- Wheel and steering encoders
- Fiber optic gyro
- Pitch and roll sensors

These robots either travel directly to unexploded submunition targets and perform the desired operation (BIP or PUCA) or perform pattern searches of the designated areas. Each robot autonomously navigates to a target location by using its local navigation system to avoid obstacles and other robots along the way. Recently, stereovision and ultrasonic sensors were added for obstacle detection/avoidance and for traffic management improvement. Once the target has been detected and localized, the robot either places a countercharge on the target for later remote initiation, or picks-up the target and deposits it at a common disposal point. This process is repeated until EOD technicians override and control a particular robot, as needed. Because the BUGS system operates semi-autonomously, one or two EOD operators will be able to command a large swarm to achieve efficiency in clearance operations.

4. TEST AND EVALUATION

The purpose of Test and Evaluation (T&E) is to do a data comparison of two robotic systems and evaluate their performances for feeding into an AOA study. The systems are being tested and evaluated to determine BUGS performance.

1) Compare systems performance and operation capability. (Simple / Highly Functional)
2) Check System Supportability and ease of use. (Training and Maintainability)
3) Evaluate Operational / Technical risks (Performance / Time / Cost)
4) Evaluate system approach and cooperative behavior.

4.1   Preliminary Test Results

Figure 3. BUGS Test Area

The preliminary first phase of PUCA operations for the highly functional system have been successfully tested at both the GDRS and NAVEODTECHDIV locations.

With the direct search, the location and position of the targets have been identified by another system. The coordinates of the target location are fed into the OCU for system traffic management. A group of five vehicles was directed to 25 known UXO locations. The group of robots was able to successfully detect and pick-up the ordnance with minimal operator intervention in an average time of 50 minutes.

With the pattern search, without known target location, one vehicle was sent to a 50 x 50 feet area to performing a full area search. 100% coverage was achieved with 50 percent overlap between tracks to allow for navigation errors. The robot was able to successfully recover 100 percent (25 UXO) in 1hour and 10 minutes on average.

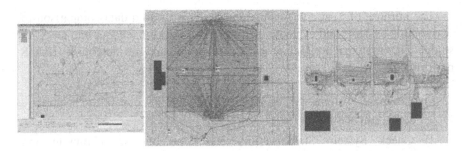

Figure 4. Direct Search and Pattern Search

For the random system, the data shows that even with a randomly searching approach, the data is still consistent and proportional. Throughout seven tests, the robots show that they do cover approximately the same amount of area in a given period of time. A single robot can repeatedly cover 25%~30% of the entire area even if its area of coverage is different from time to time. The data consistently shows that even though the robots run randomly covering different areas, the percentage of area coverage is proportionally the

3/15/00

Time (mins)	# Grids Covered	% Area Covered (# Covered / Total)	# Grids Covered	% Area Covered (# Covered / Total)	# Grids Covered	% Area Covered (# Covered / Total)
5	69	0.1104	60	0.096	62	0.0992
10	136	0.2174	129	0.2064	122	0.1952
15	192	0.3072	174	0.2784	162	0.2592
20	236	0.3776	216	0.3456	197	0.3152
25	273	0.4368	250	0.4	245	0.392
30	317	0.5072	283	0.4528	282	0.4512
35	356	0.568				
40	387	0.6192				
45	408	0.6528				
50	417	0.6672				
55	422	0.6752				
60	435	0.696				

Figure 5. Area Coverage Test Result

same, with a small deviation. Another encouraging data set shows that by doubling the number of robots, the percentage of area covered doubles, and the time for the same area coverage is approximately cut in half.

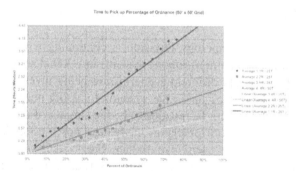

Figure 6. 50ft x 50 ft area

Another data set shows that a single robot system, operating within a 50x50 foot area, is capable of clearing 80% of the targets in four hours. By doubling the number of robots, 80% of the targets are cleared within two hours, and by quadrupling the number of robots the time is cut by one-fourth compared to a single robot performance. Also, by doubling the number of targets within the same area, the 80% clearance time only increases slightly. (Figure 6) Further testing shows, by quadrupling the test site to 100x100 feet, the 80% clearance rate still holds for the multiple robot system (Figure 7).

Further testing will be carried out to ascertain the performance of both systems. In this way, the system Performance / Time / Cost can be evaluated.

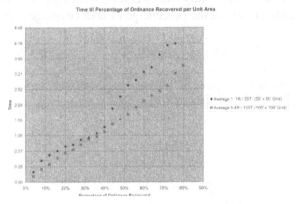

Time til Percentage of Ordnance Recovered per Unit Area

Figure 7. 50ft x 50 ft vs. 100ft x 100ft

5.   SUMMARY AND CONCLUSIONS

The Naval EOD Technology Division is continuing with its testing and analyzing multiple autonomous robotic systems in UXO clearance. Test plans are being carried out and different scenarios are being investigated, that will aid in the determination of field able systems for EOD needs. The random search configuration system is intended to be the simpler of the two, primarily because it does not require precision navigation nor a-priori knowledge of target locations. This system can reduce the system operation cost by minimizing the need for precise navigation, assistance assets, and operators. The pattern search system is more complex, but has more capabilities in performing comprehensive area searches, creating the UXO clearance map dynamically, and allowing the user capability to interface with the system.

The key element to the success of the multiple robot system is based on its performance, simplicity, and low cost. The multiple robots (BUGS) approach in UXO and mine clearance is an excellent way of distributing assets while capable of removing the EOD personnel from danger. It is believed the BUG system is inexpensive, simple, and capable of accomplishing the mission faster and cheaper, based on its new improved conventional clearance techniques.

Acknowledgments

The authors would like to acknowledge Michael Toscano and the Office of the Secretary of Defense's Joint Robotics Program, and Dr. Thomas Swean and the Office of Naval Research for sponsorship and guidance.

References

Greiner, H., Angle, C. M., Jones, I. J., Shectman, A., and Myers, R. (1996). Enabling Techniques for Swarm Coverage Approaches. In *Proceedings of Second Symposium on Technology and the Mine Problem*, Monterey, CA.

Jenkins, D. A. (1995). 'BUGS' Basic Unexploded Ordnance Gathering System: Effectiveness of Small Cheap Robotics. Naval Postgraduate School, Monterey, CA.

Mataric, M.J. (1994). Interaction and Intelligent Behavior. Technical Report AIM-1495, MIT Artificial Intelligence Laboratory.

McLurkin, J. D. (1997). Using Cooperative Robots for Explosive Ordnance Disposal. Massachusetts Institute of Technology research for NAVEODTECHDIV.

Nguyen, T., Debolt, C., and Nguyen, TB. (1999). Swarms of Small Robots for Autonomous UXO Clearance, the BUGS Project. In *Proceedings of Autonomous Unmanned System AUVSI'99*, Baltimore, MD.

O'Donnell, C., Freed, C., and Nguyen, T. (1995). BUGS: An Autonomous 'Basic UXO Gathering System' Approach in Submunition and Minefield Neutralization and Countermeasures. In *Proceedings of Autonomous Vehicles in Mine Countermeasures Symposium*, pages 8-23-8-31, Monterey, CA.

II

DISTRIBUTED SURVEILLANCE

PROGRAMMING AND CONTROLLING THE OPERATIONS OF A TEAM OF MINIATURE ROBOTS *

Paul E. Rybski, Sascha A. Stoeter,
Maria Gini, Nikolaos Papanikolopoulos
Center for Distributed Robotics, Department of Computer Science and Engineering
University of Minnesota, Minneapolis, MN 55455
{ rybski,stoeter,gini,npapas } @cs.umn.edu

Abstract   We describe a software architecture used to control the operations of a group
of miniature mobile robots called Scouts. Due to their small size, the Scouts
rely on a proxy processing scheme where they receive commands and transmit
sensor information over RF channels to a controlling workstation. Because the
bandwidth of these channels is limited, a scheduling system has been developed
that allows the robots to share the bandwidth. Experimental results are described.

Keywords: Miniature robots, distributed sensing, resource sharing

1.   INTRODUCTION

Tasks with multiple robots require a software framework in which behaviors
can be easily integrated, and in which access to resources can be scheduled
and managed by the controlling software without much user intervention. We
have developed a distributed software system for controlling a group of small,
mobile robots which have extremely limited on-board computing capabilities.
These robots, called Scouts, are completely reliant upon a proxy processing
scheme for all their computing needs, including the digitizing and processing
of the video data they broadcast over a fixed-frequency analog radio link.

The communication channels the Scouts use to send and receive information are very limited in power and throughput. As a result, access to these
channels must be explicitly scheduled so that the demand for them can be met

---

*Material based upon work supported by the Defense Advanced Research Projects Agency, Microsystems
Technology Office (Distributed Robotics), ARPA Order No. G155, Program Code No. 8H20, issued by
DARPA/CMD under Contract #MDA972-98-C-0008.

A.C. Schultz and L.E. Parker (eds.), Multi-Robot Systems: From Swarms to Intelligent Automata, 65-72.

while maintaining the integrity of the system's operation. The Scout control architecture has been developed to take these factors into account.

We present experimental results on a distributed surveillance task in which multiple Scouts automatically position themselves in an area and watch for motion. We discuss how the limited communication bandwidth affects robot performance.

2.   SCOUT ROBOTS

Scouts are miniature (11.5 cm in length and 4 cm in diameter) robotic systems designed for surveillance and reconnaissance tasks (Rybski et al., 2000). They have a video camera which they use to transmit images to a remote source. They communicate over a packetized RF communications link using an ad-hoc networking protocol. Due to the Scout's limited volume and power constraints, the two on-board computers are only powerful enough to handle communications and actuator controls. All decisions and sensor interpretations are done on an off-board workstation or by a human teleoperator. Figure 1 shows a group of the robots.

Figure 1.   The fleet of Scout robots.

Video data is broadcast over a fixed-frequency analog radio link and must be captured by a video receiver and fed into a framegrabber for digitizing. Because the video is a continuous analog stream, only one robot can broadcast on a given frequency at a time. Signals from multiple robots transmitting on the same frequency disrupt each other and become useless.

The RF limitations of the Scout pose a couple of fundamental difficulties when trying to control several of them. First, the command radio has a fixed bandwidth. This limits the number of commands it can transmit per second, and therefore limits the number of Scouts that can be controlled simultaneously. Currently, we operate on a single carrier frequency, with a command throughput of 20-30 packets/second, which is sufficient to control 4 to 5 Scouts.

The most important problem is that there are not enough frequencies available in commercial off the shelf video transmitters to allow for a large number of simultaneous analog transmissions. With the current Scout hardware we have only two video frequencies. As a result, video from more than two robots requires interleaving the time each robot's transmitter is on. Thus, an automated scheduling system is required.

3. DYNAMIC RESOURCE ALLOCATION

We have designed a distributed software architecture (Stoeter et al., 2000), which dynamically coordinates hardware resources across a network of computers and shares them between client processes.

Access to physical hardware is controlled through components (software processes) called **Resource Controllers** (or RCs). If a decision process needs to use a resource, it must be granted access to its RC. Resources that can only be managed by having simultaneous access to groups of RCs are handled by a second layer components called **Aggregate Resource Controllers** (or ARC).

In order for a process to control a Scout, several physical resources are required. First, a robot not currently in use by another process must be selected. Next, a command radio with the capacity to handle the demands of the process is needed. If the Scout is to transmit video, exclusive access to a fixed video frequency is required, together with a framegrabber connected to a tuned video receiver.

Each instance of these four resources is managed by its own RC. In Figure 2 solid lines indicate which RCs the ARCs currently have access to. Dashed lines indicate RCs which are exclusive access only and can only support control from a single ARC. The **Radio RC** is an exception to this, as it is a sharable RC. Since ARC-2 has access to all of its RCs, it can run. ARC-1 cannot run because it is waiting on two RCs.

Access to RCs must be scheduled when there are not enough RCs to satisfy the requirements of the ARCs. A centralized process called the RESOURCE CONTROLLER MANAGER maintains a master schedule of all active ARCs and grants access to each of their RCs when it is their turn to run. When requesting access to a set of RCs, an ARC must specify a minimum amount of time that it must run to get any useful work done.

The RESOURCE CONTROLLER MANAGER's scheduling algorithm tries to grant simultaneous access to as many ARCs as possible. The ARCs that have

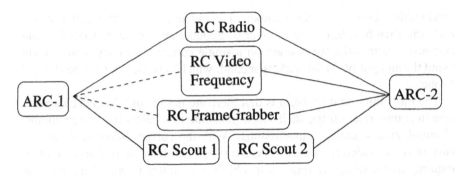

Figure 2. An example of how a decision process controls RCs by connecting to a single ARC.

some RCs in common are examined to determine which ARCs can operate in parallel and which are mutually exclusive. ARCs which request a non-sharable RC cannot run at the same time and must break their total operating time into slices. ARCs which have a sharable RC in common may be able to run simultaneously, assuming that the capacity requests for that sharable RC do not exceed its total capacity. Once the ARC schedule has been constructed, the RESOURCE CONTROLLER MANAGER executes it and takes care of notifying the RCs which ARC they should talk to at any given point in the schedule.

4.   EXPERIMENTAL RESULTS

In the experiments the Scouts are used in a distributed surveillance task where they are deployed into an area and watch for motion. This is useful in situations where it is impractical to place fixed cameras because of difficulties relating to power, portability, or even the safety of the operator.

Several simple behaviors have been implemented to do the task. All the behaviors use the video camera, which currently is the only environmental sensor available to the Scout. These behaviors include Locate Goal which rotates the Scout in place while searching the area around it for a target area of interest, Drive Toward Goal which visually servos the robot to an area of interest, Handle Collisions which helps disengage the Scout from an obstacle, and Detect Motion in which the Scout robot reports whether something in its field of view is moving.

To test the ability of the Scouts to operate in a real-world environment, a test course was set up in our lab using chairs, lab benches, cabinets, boxes, and miscellaneous other materials. The goal of each Scout was to find a suitable dark hiding place, move there, turn around to face a lighted area of the room, and watch for motion.

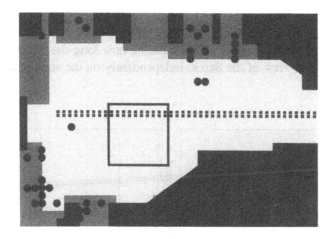

Figure 3.   A top-down view of the room where the experiments were conducted. White areas are open space, gray areas are hiding spaces, and black areas are obstacles. The square outline in the center shows where the Scouts were started, the dotted line indicates the path of the moving target, and the dots are the hiding positions of the Scouts.

The environment, shown in Figure 3, is 6.1 by 4.2 m and has a number of secluded areas in which the Scouts could hide. The Scouts were started at the center of the room and were pointed at one of 8 possible orientations. Both the position and orientation were chosen from a uniform random distribution.

The moving target the Scouts had to detect was a commercial mobile robot, chosen for its ability to repeatedly travel over a path at a constant speed. The target moved at a speed of approximately 570 mm/s and traversed the the room in 8.5 seconds on average. Once it had moved 16 feet into the room, it turned around and moved back out again. With a 4 second average turn time, the average time the target was in the room was 21 seconds.

When Scouts shared a single video frequency, only one Scout at a time could access the video frequency. Access was allocated and scheduled by the RESOURCE CONTROLLER MANAGER. The amount of time each behavior could use the video frequency was set to 10 seconds, 3 seconds of which were needed every time for the video-transmitter to warm up, so leaving 7 seconds for useful work.

Four different cases were tested: (1) a single Scout using a single video frequency, (2) two Scouts sharing a single video frequency, (3) two Scouts using two different video frequencies, (4) four Scouts sharing two different video frequencies. We run a total of 200 trials, with different hiding positions and number of scouts.

To evaluate the motion detection abilities of the Scouts and to determine the effect of sharing the video frequency, the actual time the target was seen (shown in Figure. 5) was compared to the potential time that the target could have been seen given the Scout positions (shown in Fig. 4). This potential time was calculated by analytically computing how long the target would be within the field of view of the Scout, independently on the state of activity of the Scout.

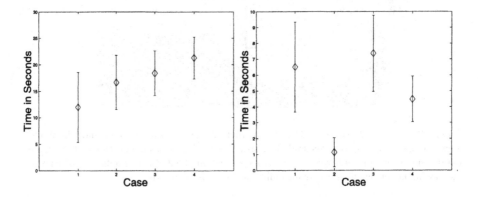

Figure 4. The potential time (in seconds) the Scouts could have been seen the moving robot. (1) one Scout, (2) two Scouts on a single frequency, (3) two Scouts on two different frequencies and (4) four Scouts on two different frequencies. Plots show means and standard deviations.

Figure 5. The actual time (in seconds) the Scouts detected motion. (1) one Scout, (2) two Scouts on a single frequency, (3) two Scouts on two different frequencies and (4) four Scouts on two different frequencies. The plots show means and standard deviations.

One Scout on a single frequency had a much higher success rate than two Scouts on a single frequency. This was expected because when the Scouts had to take turns with the video frequency, they could easily miss the target. The shorter the time the target was in the field of view, the smaller was the opportunity for the Scout to detect it, even when there was no swapping because a single frequency was used. Using a larger number of Scouts increased the viewable area traversed by the target and the time that the target was in view, and decreased the variances.

The area viewed by four Scouts was significantly greater than the areas viewed in the other cases, but not by a factor of four over that viewed by one Scout nor by a factor of two over that viewed by two Scouts. The environment was such that there was usually a great deal of overlap in the areas viewed by individual Scouts. Redundancy was probably not as useful in this environment, but would probably be more effective in larger or more segmented environments. More details on the experiments are in (Rybski et al., 2001).

5. RELATED WORK

Due to the small size, most miniature robots use proxy processing, as in Inaba *et al.* (Inaba et al., 1996), and communicate via a wireless link with the unit where the computation is done. This becomes a problem when the bandwidth is limited, as in the case of our Scouts. Because of their limited size, not only all processing is done off-board but also the communication is limited to a few communications channels.

A number of software architectures have been proposed for multiple robots, many of them described in (Kortenkamp et al., 1998). Our architecture has some similarities with ALLIANCE (Parker, 1998) and CAMPOUT (Pirjanian et al., 2000). The major difference is that we focus on resource allocation and dynamic scheduling, while other architectures are designed for more complex behavior fusion.

Resource allocation and dynamic scheduling are essential to ensure robust execution. Our work focuses on dynamic allocation of resources at execution time, as opposed to analyzing resource requests off-line, as in (Atkins et al., 2001; Durfee, 1999), and modifying the plans when requests cannot be satisfied. Our approach is specially suited to unpredictable environments, where resources have to be allocated in a dynamic way that cannot be predicted in advance. We rely on the wide body of algorithms that exists in the area of real-time scheduling (Stankovic et al., 1998) and load balancing (Cybenko, 1989).

6. SUMMARY AND FUTURE WORK

An essential feature of the distributed software control architecture we presented is the ability to dynamically schedule access to physical resources, such as communication channels and framegrabbers, that have to be shared by multiple robots.

We have also presented system issues related to the control of multiple robots over a low bandwidth communications channel. Experimental results illustrating the ability of the Scout to position itself in a location ideal for detecting motion and the ability to detect motion have been shown. Future work is planned to allow the Scouts to make use of additional sensor interpretation algorithms for more complex environmental navigation. Ultimately, we hope to have the Scouts construct a rudimentary topological map of their surroundings, allowing other robots or humans to benefit from their explorations.

We believe that a combination of intelligent scheduling and more flexible hardware will allow a larger number of Scout robots to operate simultaneously in an effective manner.

References

Atkins, E. M., Abdelzaher, T. F., Shin, K. G., and Durfee, E. H. (2001). Planning and resource allocation for hard real-time, fault-tolerant plan execution. *Autonomous Agents and Multi-Agent Systems*, 4(1/2):57–78.

Cybenko, G. (1989). Dynamic load balancing for distributed memory multiprocessors. *Journal of Parallel Distributed Computing*, 7(2):279–301.

Durfee, E. H. (1999). Distributed continual planning for unmanned ground vehicle teams. *AI Magazine*, 20(4):55–61.

Inaba, M., Kagami, S., Kanechiro, F., Takeda, K., Tetsushi, O., and Inoue, H. (1996). Vision-based adaptive and interactive behaviors in mechanical animals using the remote-brained approach. *Robotics and Autonomous Systems*, 17:35–52.

Kortenkamp, D., Bonasso, R. P., and Murphy, R. (1998). *Artificial Intelligence and Mobile Robots*. AAAI Press/MIT Press.

Parker, L. E. (1998). ALLIANCE: An architecture for fault tolerant multi-robot cooperation. *IEEE Trans. on Robotics and Automation*, 14(2):220–240.

Pirjanian, P., Huntsberger, T., Trebi-Ollennu, A., Aghazarian, H., Das, H., Joshi, S., and Schenker, P. (2000). CAMPOUT: a control architecture for multirobot planetary outposts. In *Proc. SPIE Conf. Sensor Fusion and Decentralized Control in Robotic Systems III*.

Rybski, P. E., Papanikolopoulos, N., Stoeter, S. A., Krantz, D. G., Yesin, K. B., Gini, M., Voyles, R., Hougen, D. F., Nelson, B., and Erickson, M. D. (2000). Enlisting rangers and scouts for reconnaissance and surveillance. *IEEE Robotics and Automation Magazine*, 7(4):14–24.

Rybski, P. E., Stoeter, S. A., Gini, M., Hougen, D. F., and Papanikolopoulos, N. (2001). Performance of a distributed robotic system using shared communications channels. Technical Report 01-031, Computer Science and Engineering Department, University of Minnesota.

Stankovic, J., Spuri, M., Ramamritham, K., and Buttazzo, G. (1998). *Deadline Scheduling For Real-Time Systems: EDF and Related Algorithms*. Kluwer Academic Publishers, Boston.

Stoeter, S. A., Rybski, P. E., Erickson, M. D., Gini, M., Hougen, D. F., Krantz, D. G., Papanikolopoulos, N., and Wyman, M. (2000). A robot team for exploration and surveillance: Design and architecture. In *Proc. of the Int'l Conf. on Intelligent Autonomous Systems*, pages 767–774, Venice, Italy.

AUTONOMOUS FLYING VEHICLE RESEARCH AT THE UNIVERSITY OF SOUTHERN CALIFORNIA

Srikanth Saripalli, David J. Naffin and Gaurav S. Sukhatme
Computer Science Department
University of Southern California, Los Angeles
California, USA
srik,dnaffin,gaurav@robotics.usc.edu

Abstract   This paper outlines research on Autonomous Flying Vehicles at the University of Southern California (USC). We are particularly interested in control strategies for autonomous vehicles to perform difficult tasks such as autonomous landing and trajectory following. In addition we are starting research on cooperative algorithms for formation flight with a group of autonomous flying robots. In particular, we present the design and behavior-based control architecture for an autonomous flying vehicle (AFV), for vision-based landing. We use vision for precise target detection and recognition as well as combination of vision and GPS for navigation. An outline of the reference design of an autonomous helicopter for research in formation flying is also presented. A discussion of exemplar algorithms used for controlling multiple robots in formation is presented.

Keywords: Autonomous Flying Vehicle, Formations, Vision-based landing, Multiple Robot Coordination

1.   INTRODUCTION

A basic requirement for autonomous flying vehicles (AFVs) is robust autonomous flight, for which *autonomous landing* is a crucial capability. The problem of autonomous landing is particularly difficult because the inherent instability of the helicopter near the ground (Shakernia et al., 1999). Also since the dynamics of a helicopter are non-linear only an approximate model of the helicopter can be constructed (Conway, 1995). (Garcia-Pardo et al., 2000) discusses a vision-based solution to safe landing in unstructured terrain where the key problem is for the onboard vision system to detect a suitable place to land, without the aid of a structured landmark such as a helipad. (Sinopoli

A.C. Schultz and L.E. Parker (eds.), Multi-Robot Systems: From Swarms to Intelligent Automata, 73-80.
© 2002 *Kluwer Academic Publishers. Printed in the Netherlands.*

et al., 2001) have demonstrated tracking of a landing pad based on vision but have not shown landing as such.

At the same time very little research has been performed in the area of co-ordination of multiple AFVs. AFV systems tend to be expensive and complex and therefore just having one system is considered a major accomplishment. Why fly in formations? Beside the challenge of accurately controlling multiple robots, there is the added benefit of having the coordination of multiple sets of sensors. Nature favors animals that have the ability to form formations such as flocks of birds or schools of fish. Animals that can combine their sensing abilities have shown to better avoid predators and efficiently forage for food (Balch and Arkin, 1998). Both the Air Force and NASA have identified autonomous formation of spacecraft as key technological milestones for the 21st century (Beard et al., 2000). Applications of spaced based autonomous vehicles range from ground surveillance to interferometry experiments.

Our research spans from developing better sensing capabilities to algorithms for multi-robot formations. The AVATAR (Autonomous Vehicle Aerial Tracking And Reconnaissance) project has been an ongoing research project at USC for the past ten years. The primary focus of the AVATAR project has been developing greater sensing and control capabilities for a single UAV, while at the same time cooperation and coordination with ground based robots. For a description of the research regarding multi-robot coordination between aerial and ground vehicles the reader is referred to (Sukhatme et al., 2001). The project has recently reached a major milestone; it is the first UAV that has landed autonomously under vision-based control. The RAPTOR (Research on Aerially Precise Teams of Robots) project at USC has begun investigating multi-robot formation control. In particular we plan on utilizing small electric-powered radio-controlled (R/C) model helicopters as a mechanical chassis for our design. Relative localization of each robot will be accomplished using only local sensing (CMOS camera). In contrast to global localization techniques (GPS), each robot will only have knowledge of it relative location with respects to one or more of its neighbors. This will be beneficial in environments where GPS is not available (i.e. indoors or between skyscrapers).

In this paper we present two different facets of our research. The first part of the paper describes the use of vision coupled with Inertial Navigational System to perform autonomous landing. The second part of the paper describes our ongoing research in the field of formation flying using AFVs.

2.   THE AVATAR TESTBED

Our experimental testbed AVATAR (Autonomous Vehicle Aerial Tracking And Reconnaissance) (Montgomery, 1999) is a gas-powered radio-controlled model helicopter fitted with a PC-104 stack augmented with several sensors

Figure 1. AVATAR (Autonomous Vehicle Aerial Tracking And Reconnaissance)

(Figure 1). A Novatel RT-20 DGPS system provides positional accuracy of 20cm CEP (Circular Error Probable, i.e. the radius of a circle, centered at the true location of a receiver antenna, that contains 50% of the individual position measurements made using a particular navigational system). A Boeing CMIGTS-II INS unit with three axis accelerometers and three-axis gyroscopes provides the state information to the on-board computer. The helicopter is equipped with a color CCD camera and an ultrasonic sonar. The ground station is a laptop that is used to send high-level control commands and differential GPS corrections to the helicopter. Communication with the ground station is carried via 2.4 Ghz wireless Ethernet and 1.8Ghz wireless video.

3. CONTROL ARCHITECTURE

The AVATAR is controlled using a hierarchical behavior-based control architecture. Briefly, a behavior-based controller (Mataric, 1997) partitions the control problem into a set of loosely coupled behaviors. Each behavior is responsible for a particular task. The behaviors act in parallel to achieve the overall goal. Low-level behaviors are responsible for robot functions requiring quick response while higher-level behaviors meet less time critical needs. The behavior-based control architecture used for the AVATAR is shown in Figure 2.

At the lowest level the robot has a set of reflex behaviors that maintain stability by holding the craft in a hover condition. A detailed description of the architecture can be found in (Montgomery, 1999). A key advantage of such a

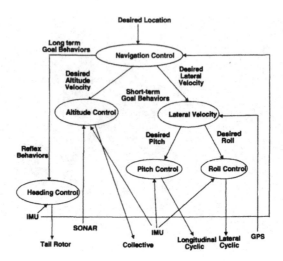

Figure 2.　　Behavior Based Controller

control algorithm is to build complex behaviors on top of the existing low level behaviors.

The low-level and short-term goal behaviors *roll, pitch, heading, altitude and lateral control* behaviors are implemented with proportional controllers.The long-term goal behavior *navigation control* is responsible for overall task planning and implementation. If the heading error is small, the *navigation control* behavior gives desired lateral velocities to the *lateral velocity* behavior. If the heading error is large, the *heading control* behavior is commanded to align the helicopter with the goal while maintaining zero lateral velocity.

4.　　VISION-BASED STATE ESTIMATION

This section deals with the vision algorithms implemented for autonomous detection, tracking of a helipad and landing on it. The images obtained from the camera are noisy and the frame grabber is of low quality, hence we work with binary images to reduce the computational cost and increase the effectiveness of the algorithm.This is done by thresholding and filtering the image. The image obtained after thresholding and filtering may consist of objects other than the helipad. In the next step the various regions of interest are identified and labelled. The next stage involves the detection of the landing pad (in our case a Helipad) using geometric invariants. Geometric shapes possess features such as *perimeter, area, moments* that often carry sufficient information for the task of object recognition.Based on the geometric features of an object one can calculate a set of descriptors which are invariant to rotation, translation and scaling. One such class of descriptors (Hu, 1962) is based on the moments of inertia of an object. These descriptors are used for locating the helipad. For a detailed description of the algorithm the readers are referred to (Saripalli et al., 2002). After the helipad is located, the state estimation algorithm calculates the coordinates and orientation of the landing target relative to the helicopter. These state estimates are sent to the helicopter controller.

4.1 Controller for Autonomous Landing

The controller described in Section 3 is modified as below to perform vision-based landing. The *altitude control* behavior was split into three sub-behaviors, *hover-control, velocity-control and sonar-control*. The *hover-control* sub-behavior is activated when the helicopter is either flying to a goal or is hovering over a particular target. This sub-behavior is used during the object recognition and object tracking state when the helicopter should move laterally at a constant altitude. The hover controller is implemented as a proportional controller. It reads the desired GPS location and the current location and calculates the collective command to the helicopter.

Once the helipad has been located and the helicopter is aligned to the object the *velocity control* sub-behavior takes over from the *hover-control* sub-behavior. It is implemented as a PI controller. An integral term is added to reduce the steady state error. The helicopter starts to descend till reliable values are obtained from the sonar. The *sonar-control* sub-behavior takes over at this point until touchdown. This is also implemented as a PI controller. Results based on flight data from field tests show that our method is able to land the helicopter on the helipad repeatably and accurately. On an average the helicopter landed to within 31 cm position accuracy and to within $6°$ in orientation as measured from the center of helipad and its principal axis respectively. Results also show the robustness of the algorithm, which allows the helicopter to find the helipad after losing it momentarily. They show that the algorithm is capable of tracking a moving landing target and land on it, once it has stopped. In these experiments the helipad was moved a significant distance (7 m on an average). For a detailed discussion on the results and the accuracy of the controller, refer to (Saripalli et al., 2002). In the next section we describe our ongoing research on the use of autonomous flying vehicles for formation flying.

5. RAPTOR

Research on Aerially Precise Teams of Robots (RAPTOR) has developed a reference platform for research of multiple robot formations. Instead of building several large scale models such as the platform used by the AVATAR project, we have opted for a new platform based on a small electric model helicopter. The next two sections outline the design decisions that were made.

5.1   RAPTOR Mechanics and Avionics

After surveying many different R/C model helicopters, we decided on the Lite Machine's LMH-110 as our mechanical chassis. This small model helicopter is only 67 cm long. It has a 60 cm main rotor blade diameter and weights approximately 800 grams without batteries or any additional avionics. Early tests demonstrated its ability to lift more than half kilogram of payload. It is inexpensive, reliable and replacement parts are readily available from many mail order retailers and hobby shops. The platform avionics consist of several modules interconnected via a high-speed serial bus. The Inter-IC (I2C) Bus was selected as the mechanism by which each module would communicate to one another. Each module will have a dedicated micro-controller that provides the interface to the I2C bus. This will allow us to design, construct and test each module independently.

When selecting electronic components for these modules, there are three constraints that need to be taken into account. First, the overall size of the components must be kept as small as possible. Since we planned on using standard printed circuit boards that can be purchased at any Radio Shack store, we needed to "budget" the size of individual components to just those that would fit on these boards. Second, the components should be as light as possible. Third, each component should consume as little power as possible. High power consumption equates to more battery weight.

The choice of micro-controllers was a key early decision in our design phase. Ideally we wanted to use a single controller for all modules. Unfortunately, no single choice was perfect for all cases. However, we were able to limit ourselves to two choices; Atmel's Mega163 AVR controller and Scenix's SX28 processor. The Atmel Mega163 is one of the latest of highly integrated controllers from Atmel's Advanced RISC (AVR) line of micro-controllers. The Scenix SX28 is the fastest 8-bit micro-controller currently on the market. It can be clocked from four to 100 Mhz. Both controllers are programmable using standard "C" and development tools are freely available. Our current list of modules includes communications, actuator interface, inertial sensors, camera-laser-sonar interface, and a master controller.

5.2   Relative Localization

In order to maintain formation, each vehicle must know its position relative to some key location within the formation (often the location of the leader craft). While GPS receivers and global communications are one solution to this problem, there are limitations to GPS. GPS requires a direct line of sight between the receiver and at least four NAVSTAR satellites. This precludes its use indoor as well as between tall buildings. GPS receivers require as much as ten watts of power and require an antenna that is considerably bulky for small vehicles. In contrast, CMOS cameras are extremely small, light weight, and require only an illuminated area. However, a single camera is rarely able to ascertain absolute global positions. Fortunately for us, we need only relative positions to maintain a formation. Our approach to vehicle localization within a formation is vision-based. Each vehicle is equipped with a downward point-ing camera with a large field of view (FOV) lens. In addition, each vehicle is also equipped with an inexpensive laser pointer. Each laser pointer is attached to the vehicle in a manner such that it projects a signature (image) onto the surface below. The signature projected will be a plus (+) so that it can easily be distinguished from other features on the ground. Each laser pointer can be modulated to distinguish one vehicle's projection from another. This approach has several important merits. First, the 3D sensing problem has been effec-tively reduced to a 2D problem. This greatly reduces the required computation required for the vision detection system. Second, the number of sensors has been reduced to a single downward pointing camera.

5.3   Formation Control

Our research in control of formations of aerial robots, will concentrate on developing robust and scalable solutions. Initially we will concentrate on maintaining simple planar geometric shaped formations such as wedges and diamonds. These formations have long been used as benchmarks for formation studies in ground based formations (Balch and Arkin, 1998), (Fredslund and Mataric, 2001). However, unlike traditional control strategies that utilize direct sensing or global positioning, our techniques will only rely on local sensing. Later, we will investigate techniques for dynamical changing of formations and maintain formations in the presents of obstacles We plan on validating our ideas both in simulation with tens of robots as well as with a few physical robots.

For any given formation, there are two types of participants: leaders and fol-lowers. The leaders of a formation occupy the control points of the formation. The overall direction of motion and orientation of a formation is defined by the formation's leaders. The followers create line segments at specified angle with respect to the leader. A formation can be defined as one or more sets of leader/followers.

With only local sensing of projected signatures on the ground, any given robot will only "see" a small set of its neighbors. Therefore any given follower robot may not be able to see a leader and will need to maintain a heading and distance from one of the robots it can sense. Although there will be global communication between members of the formations, the bandwidth is very limited and therefore broadcast messages will be kept to a minimal.

6.   CONCLUSIONS

This paper describes ongoing research on autonomous flying vehicles at the University of Southern California. We have successfully demonstrated landing in unstructured terrain based on vision and GPS. In the future we plan to focus our attention on the problem of safe and precise landing of the helicopter in unstructured harsh 3D environments. The applications of such a system are enormous; from space exploration to target tracking and acquisition. We are starting research in multi-robot aerial vehicle formation flying. The research platform for such experiments has been presented and some initial ideas for formation flying have been proposed.

References

Balch, T. and Arkin, R. C. (1998). Behavior-based formation control for multi-robot teams. In *IEEE Transaction on Robotics and Automation*.

Beard, R. W., Lawton, J., and Hadaegh, F. Y. (2000). A coordination architecture for spacecraft formation control. In *IEEE Transaction on Control Systems Technology*.

Conway, A. R. (1995). *Autonomous Control of an Unstable Helicopter Using Carrier Phase GPS Only*. PhD thesis, Stanford University.

Fredslund, J. and Mataric, M. J. (2001). Robot formations using only local sensing and control. In *Proceedings of IEEE International Sysmposium on Computational Intelligence for Robotics and Automation (CIRA)*, Banff, Canada.

Garcia-Pardo, P. J., Sukhatme, G. S., and Montgomery, J. F. (2000). Towards vision-based safe landing for an autonomous helicopter. *Robotics and Autonomous Systems*. (accepted, to appear).

Hu, M. K. (1962). Visual pattern recognition by moment invariants. In *IRE Transactions on Information Theory*, volume IT-8, pages 179–187.

Mataric, M. J. (1997). Behavior-based control: Examples from navigation, learning and group behavior. *Journal of Experimental and Theoretical Artificial Intelligence, special issue on Software Architecture for Physical Agents*, 9(2-3):67–83.

Montgomery, J. F. (1999). *Learning Helicopter Control through 'Teaching by Showing'*. PhD thesis, University of Southern california.

Saripalli, S., Montgomery, J. F., and Sukhatme, G. S. (2002). Vision-based autonomous landing of an unmanned air vehicle. In *Proceedings of IEEE International Conference on Robotics and Automation*. (to appear).

Shakernia, O., Y.Ma, Koo, T. J., and Sastry, S. (1999). Landing an unmanned air vehicle:vision based motion estimation and non-linear control. In *Asian Journal of Control*, volume 1, pages 128–145.

Sinopoli, B., Micheli, M., Donato, G., and Koo, T. J. (2001). Vision based navigation for an unmanned aerial vehicle. In *Proceedings of IEEE International Conference on Robotics and Automation*, pages 1757–1765.

Sukhatme, G. S., Montgomery, J. F., and Vaughan, R. T. (2001). Experiments with cooperative aerial-ground robots. In *Robot Teams: From Diversity to Polymorphism*.

III

MANIPULATION

DISTRIBUTED MANIPULATION OF MULTIPLE OBJECTS USING ROPES

Bruce Donald, Larry Gariepy, and Daniela Rus
Department of Computer Science
Dartmouth College
Hanover, NH, USA[*]
rus@cs.dartmouth.edu

Abstract   This paper describes a system in which multiple robots cooperate to move multiple objects such as groups of boxes using a constrained prehensile manipulation mode, by wrapping ropes around them. The system consists of three manipulation skills: tieing ropes around objects, affecting rotations using a flossing manipulation gait, and affecting translations using a ratcheting manipulation gait. We present algorithms for these operations, a numerical analysis for the motion of groups of boxes, and experimental results.

Keywords: distributed manipulation, cooperative manipulation

1.   INTRODUCTION

We are working on distributed manipulation problems, in which two or more robots apply spatially-distributed forces to rearrange the objects in their work space; the computation is distributed among the robots so as to qualify as a distributed computation. In our paper (Donald et al., 1999) we presented the concept of distributed manipulation with ropes and demonstrated it by describing a system in which a single object can be repositioned with this technique. In this paper we describe a new class of distributed manipulation algorithms in which multiple mobile robots manipulate multiple objects at the same time, using ropes. Distributed manipulation with ropes allows us to explore the passive kinematics of ropes, when a rope is attached to the waistline of the robots. There are two interesting properties worth noting: (1) ropes cannot be actu-

---

[*]Bruce Donald is grateful to the NSF for support provided through the following grants: NSF IIS-9906790, NSF EIA-9901407, NSF 9802068, NSF CDA-9726389, NSF EIA-9818299, NSF CISE/CDA-9805548, NSF IRI-9896020, and NSF IRI-9530785. Daniela Rus is grateful to the NSF for support provided through the NSF CAREER award IRI-9624286, NSF award IRI-9714332, NSF award EIA-9901589, NSF award IIS-98-18299, and a Sloan fellowship.

A.C. Schultz and L.E. Parker (eds.), Multi-Robot Systems: From Swarms to Intelligent Automata, 83-91.

ated independently; and (2) when a rope is attached to a robot, it can *affect* the motion of the robot and also it can *effect* manipulation when the robot moves. Thus, our approach is an exploration of underactuated distributed manipulation systems based on a simple biomimetic design inspired by flagella and cilia.

Manipulating objects with ropes has several advantages over the direct manipulation of an object using a team of robots (Rus et al., 1995; Donald et al., 1997b; Donald et al., 1997a). First, the method allows for the parallel and synchronized manipulation of multiple objects. (That is, several objects can be repositioned together). Second, wrapping a rope around an object permits *constrained prehensile manipulation* with non-prehensile robots. The rope can conform to any object geometry on-line, in that the same wrapping protocol works for all geometries and there is no need to specify a model in advance. Finally, the rope can be viewed as a tool that allows the robot system to exert torques that are larger than the torque limit of each individual robot. Our notion of constrained prehensile manipulation is different than the classical definition of prehensile manipulation, which usually denotes a force closure grasp. In *constrained prehensile manipulation,* perturbations along any differential direction can be resisted (in our case, due to the taut rope) but the object can be moved only along certain lower dimensional degrees of freedom. Thus, constrained prehensile manipulation systems are *non-holonomic.*

Our results on distributed manipulation have been inspired by previous work in motion planning (Latombe, 1991), manipulation (Erdmann and Mason, 1986; Goldberg, 1993; Rus, 1999), cooperative robotics(Noreils, 1986; Parker, 1994; Yamamoto, 1994; Beckers et al., 1994), and distributed manipulation (Rus et al., 1995; Donald et al., 1997b; Donald et al., 1999; Donald et al., 2000).

2.   A DISTRIBUTED MANIPULATION SYSTEM

Consider the task whose goal is to change the pose of a a group of boxes. A team of robots can use a sequential non-prehensile strategy such as those described in (Rus et al., 1995; Donald et al., 1997b; Donald et al., 1997a) to accomplish this task: each object will be located individually. A different approach is to use a parallel constrained prehensile strategy: the objects will be repositioned simultaneously. Because the objects we wish to move are large, it is impractical to envision robots equipped with big enough grippers to do such manipulation tasks. Instead, we focus on strategies that augment the basic architecture of mobile robots with simple tools. In this section we describe an approach where the robots can achieve a grasp on collections of objects by wrapping the objects with a rope (independent of the overall geometry). We then explore cooperative strategies for manipulating a collection of objects wrapped with a rope.

2.1 Experimental Setup

Our distributed manipulation team consists of three mobile robots: two RWI B14 robots equipped with sonar, discrete contact sensors, IR, and radio modems; and one RWI Pioneer robot equipped with sonar, a radio modem, and a parallel-jaw gripper. The B14 robots are named Bonnie and Clyde and the RWI Pioneer robot is named Ben. We augmented the basic hardware of these robots by using a long rope (over 5 meters) which is tied at one end to Bonnie and at the other end to Clyde. We incorporated a tool in Ben's gripper that extends the effective height of the gripper from 2 cm. to 15 cm.

2.2 The Binding Algorithm

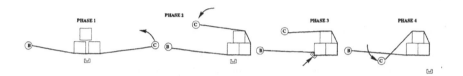

Figure 1.   The 4 phases of binding (wrapping a rope around a group of objects). The RWI robots are denoted by disks. Bonnie is denoted by B; Clyde is denoted by C. The Pioneer helper is denoted by the gripper icon. The objects are denoted by rectangles.

The goal of this skill is for a team of two robots to surround a group of objects with a rope to generate a *rope grasp*. The basic idea of the algorithm is to keep one robot stationary while controlling the other robot to move around the objects with the rope held taut. At some point, the moving robot has to cross over the rope in order to complete the grasp; this step requires coordination with the stationary robot.

We have developed a control algorithm for this skill and implemented it using two RWI B14 robots and one Pioneer robot (see Figure 1). In the first phase, the B14 robots are on the same side of the objects. Clyde, the active robot (the right robot in Figure 1) moves so as to keep the rope taut around them. Bonnie, the left robot in Figure 1, is passive and stays in place. The control for encircling the boxes is implemented as hybrid control of a compliant rotation. The robot starts with a calibration step to determine the level of power necessary to achieve rotational compliance to the box. The robot tries to move in a polygonal approximation to a circle. Due to the radial force along the rope, the robot's heading complies to remain normal to the radial force, thereby effecting a spiral. The robot continues this movement until it gets close to having to cross the rope.

The first phase ends when the encircling robot has completed a 165 degrees tour around the group of boxes. In the second phase, Clyde crosses the rope.

The algorithm requires Bonnie to drop the rope to the ground to enable Clyde to drive over the rope. A simple forward movement of Bonnie would cause the rope to drop to the ground, but it will break the grasp. Furthermore, the inverse move is not guaranteed to reestablish the grasp because the rope can slide under some box. Thus, it is important to maintain the partial grasp on the boxes during the crossover. Our approach is to use a Pioneer helper robot, that can support the rope while the passive robot gives slack.

The algorithm for the second phase starts with Ben locating the corner of the object closest to Bonnie. Ben uses its gripper to push the rope against that box, thus securing the rope grasp. Then Bonnie gives some slack in the rope. In the third phase, Clyde crosses over the rope. In the fourth phase, Bonnie tightens the rope. The tension in the rope signals the helper to let go and move away.

2.3   Ratcheting

The ratcheting algorithm translates a group of objects. However, one side effect of the ratcheting algorithm is that the group also rotates as it translates.

We have developed a control algorithm for this skill and implemented it using two RWI B14 robots. The algorithm iterates the following phases. In the first phase the active robot translates forward with velocity v_a. The effect is a transtation of the group of objects. The passive robot follows along feeling F_p and keeping the rope taut. This action causes the robot to translate. When this phase is executed for the first time, it terminates with the passive robot sensing contact with the object. This is called the *calibration step*. During a calibration step, the passive robot measures the approximate distance to the objects. In subsequent repetitions of this phase, the passive robot uses dead reckoning to decide when to stop.

In the second phase the robots decrease the tension in the rope and translate together in the opposite direction. The objects do not move in this phase and the rope slips against them. The effect of this phase is to re-grasp the objects so as to move the active robot closer to the objects.

2.4   Flossing

The flossing algorithm reorients a group of objects. The task itself does not actually effect a pure rotation, but also translates the group slightly as a side effect. To demonstrate reorientation in either direction, we have the robots perform a kind of tug-o-war over the group of boxes. The algorithm for this task is symmetric with respect to the two robots. They simply take turns towing the group (and each other), and rotating the group back-and-forth. The robots alternate between the active and passive roles in the algorithm. We use a stop signal just like the one in the ratcheting experiment to allow the passive robot to signal the active robot that it is time to switch roles.

3. ANALYSIS

In our lab, we implemented and experimented with the three manipulation primitives described in the previous section. We observed experimentally that while wrapping and manipulating groups of boxes, the boxes almost always settled into an arrangement where each box was flush (i.e. touching face to face) against its neighbors. This property leads to simplifications in the manipulation and wrapping operations. In this section we analytically explore this observation. We ask: for what class of box geometries does wrapping lead to a final configuration where each box is flush against its neighbors?

Our approach is to calculate the perimeter of the convex hull of the box arrangement and view it as a potential function. By pulling on the rope during the wrapping operation, we reduce this potential energy and force the boxes into a "lower energy state".

Notation: θ_2, θ_3–orientations of Box 2 and Box 3.

p_2, p_3–positions of Box 2 and Box 3.

$C^{\text{Box } i}_{\text{Box } j, \theta_j}$–configuration space obstacle

of Box i w.r.t. Box j oriented at angle θ_j.

Input: Geometries of Box 1, Box 2, and Box 3.

Analyze-Config: Let θ_2 loop from 0 to π.

Compute $C^{\text{Box } 1}_{\text{Box } 2, \theta_2}$.

Let p_2 loop over the perimeter of Box 1.

Let θ_3 loop from 0 to π.

Compute $C^{\text{Box } 1}_{\text{Box } 3, \theta_3}$ and $C^{\text{Box } 2}_{\text{Box } 3, \theta_3}$.

Merge $C^{\text{Box } 1}_{\text{Box } 3, \theta_3}$ and $C^{\text{Box } 2}_{\text{Box } 3, \theta_3}$.

Let p_3 loop over the (joint) perimeter of

Box 1 and Box 2.

Output: Perimeter of convex hull of

Box 1, Box 2, and Box 3.

Figure 2. Constructing Box Arrangements.

Ideally, we would like to obtain closed form solutions for arrangements of any number of boxes that minimize their perimeter. In the case of two boxes, these local minima correspond to box arrangements where the boxes have edge-edge contacts (Donald et al., 2000). However, it is not clear whether or not local minima of multiple box systems contain only edge-edge contacts. For example, with three boxes we can imagine a "jammed" configuration where one box is stuck in the corner defined by the other two boxes. Because the closed form solutions are rather involved even for the case of two boxes, we present a numerical approach to modeling three boxes that can be generalized to an arbitrary number of boxes.

Figure 2 shows the numerical algorithm that analyzes arrangements of three boxes whose geometries are known. With fixed box geometries, there are 4 variables to consider. We assume, without loss of generality, that the first box is centered at the origin, with sides parallel to the x and y axes and that all the boxes are touching. The positions and orientations of the other two boxes are (one-dimensional) free variables. The set of all possible configurations of three boxes forms a four dimensional manifold. We can define a positive real-valued function on this manifold that is the perimeter of the convex hull of the three boxes.

The algorithm works by assigning values to the free variables one at a time, using discretization and enumeration. The first loop chooses an orientation θ_2 for box 2. We now view box 1 as a configuration space obstacle for box 2 in order to position box 2 with respect to box 1. The second loop runs over the perimeter of this configuration space obstacle (using increments determined by a position resolution parameter), and generates values for p_2.

The same method can be used to add box 3 to the arrangement. Thus, the third loop chooses an orientation θ_3 for box 3 and computes its configuration space. The configuration space obstacle for the union of box 1 and box 2 is computed as the union of the configuration space obstacle for box 1 and the configuration space obstacle for box 2. This is a non-convex object. The algorithm computes the perimeter of the convex hull of this configuration space obstacle (a function of θ_2, p_2, and θ_3), and the fourth loop chooses a position p_3 along the perimeter of this configuration space obstacle for box 3. Having completely defined an arrangement of the three boxes, we can compute its perimeter and identify the smallest perimeter value.

4.   EXPERIMENTS

We have implemented the algorithm outlined in Figure 2. Since the perimeter of a three-box arrangement is a function $s : R^4 \longrightarrow R$, we cannot graphically depict where the local minima of the function occur. However, we can examine s along 2-dimensional slices, generally by fixing the position and orientation of Box 2 (p_2 and θ_2), and letting Box 3 move freely. Figure 3 shows sample potential surfaces computed in this manner.

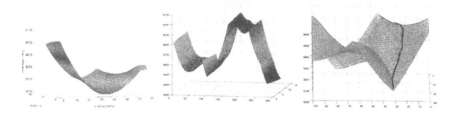

Figure 3. Left: A two-dimensional slice of the potential function for an arrangement of three boxes. Center: Wide range plot. Right: A potential plot with a highlighted simulation path

Task	Successes	Tries	Reliability
Flossing	17	18	94.4 %
Ratcheting	13	20	65 %

Figure 4. This table contains reliability data for our experiments for manipulation of three boxes by flossing and ratcheting.

Figure 3(Left) shows details for how the perimeter function changes with respect to p_3 and θ_3, and Figure 3(Center) shows a more global view of how perimeter changes with respect to p_3.

In Figure 3(Right), we demonstrate how the box arrangements depicted in Figure 5 correspond to the potential surface drawn with the simulation program. The initial snapshot of Figure 5 corresponds to the upper end of the path in Figure 3. As the box rotates and falls into place in the corner, the configuration moves down the path and into a valley, eventually stopping at a local minimum.

In our lab, we built a distributed manipulation system using ropes by implementing the manipulation primitives described in Section 2 using the robots described in Section 2.1. The RWI robots are programmed in MOBOT-SCHEME (Rees and Donald, 1992) and the helper Pioneer robot is programmed in C++.

Figure 4 shows reliability data for some recent experiments run with this system. The failures in flossing and ratcheting were due to signaling errors. Occasionally, the passive robot stops prematurely. We believe that these errors are due to low battery levels during those specific experiments, but more investigation is necessary.

We also have run many binding experiments and compared the arrangements obtained in the physical system to the simulation predictions. Figure 5 shows some snapshots from a typical run, for corresponding box geometries and initial arrangements. Since the physical experiments are subject to frictional effects that are not modeled by our simulator, we note that we sometimes observe jammed arrangements (in physical equilibrium) that do not correspond to local minima. However, in most experiments, boxes that are set up in general position very rarely achieve this state.

Figure 5.   Four snapshots from a typical wrapping run.

5.   SUMMARY

In this paper we described the constrained prehensile manipulation of multiple objects bound by a rope. We presented three algorithms for the distributed manipulation of a bound group of objects: binding (tieing a rope around a group of objects), flossing (effecting rotations of a group of bound objects), and ratcheting (effecting translations of a group of bound objects). Our algorithms use a team of three robots that do not communicate directly but use the tension in the rope as a signaling mechanism. We presented an analytical and a numerical analysis for the final configuration of a group of bound objects after wrapping. We implemented the three manipulation skills and conducted physical experiments. We also implemented the numerical solution and examined the accuracy of the numerical predictions against physical experiments.

References

Beckers, R., Holland, O., and Deneubourg, J. (1994). From local actions to global tasks: stigmergy and collective robotics. In Brooks, R. and Maes, P., editors, *Artificial Life IV*. MIT Press.

Donald, B., Gariepy, L., and Rus, D. (1999). Experiments in constrained prehensile manipulation: distributed manipulation with ropes. In Corke, P., editor, *Experimental Robotics VI*. Springer Verlag.

Donald, B., Gariepy, L., and Rus, D. (2000). Distributed manipulation of multiple objects with ropes. *Proceedings of the 2000 IEEE International Conference on Robotics and Automation*.

Donald, B., Jennings, J., and Rus, D. (1997a). Information invariants for cooperating autonomous mobile robots. *International Journal of Robotics Research*, 16:673–702.

Donald, B., Jennings, J., and Rus, D. (1997b). Minimalism + distribution = supermodularity. *Journal of Experimental and Theoretical Artificial Intelligence*, 9:293–321.

Erdmann, M. and Mason, M. (1986). An exploration of sensorless manipulation. *Proceedings of the 1986 IEEE International Conference on Robotics and Automation*.

Goldberg, K. (1993). Orienting polygonal parts without sensing. *Algorithmica*, 10:201–225.

Latombe, J.-C. (1991). *Robot Motion Planning*. Kluwer Academic Press.

Noreils, F. (1986). Toward a robot architecture integrating cooperation between mobile robots: Application to indoor environment. *International Journal of Robotics Research*, 12.

Parker, L. (1994). *Heterogeneous Multi-Robot Cooperation*. PhD thesis, MIT EECS Department, Boston, MA.

Rees, J. and Donald, B. (1992). Program mobile robots in scheme. *Proceedings of the 1992 IEEE International Conference on Robotics and Automation*.

Rus, D. (1999). In-hand dexterous manipulation of piecewise-smooth objects. *International Journal of Robotics Research*, 18:355–381.

Rus, D., Donald, B., and Jennings, J. (1995). Moving furniture with teams of autonomous robots. *Proceedings of the 1995 Conference on Intelligent Robot Systems*.

Yamamoto, Y. (1994). Coordinated control of a mobile manipulator. Technical Report MS-CIS-94-12 / GRASP LAB 372, The University of Pennsylvania.

A DISTRIBUTED MULTI-ROBOT SYSTEM FOR COOPERATIVE MANIPULATION

Aveek Das, John Spletzer, Vijay Kumar, and Camillo Taylor

GRASP Laboratory

University of Pennsylvania, Philadelpia, PA 19104, USA

aveek, spletzer, kumar, cjtaylor@grasp.cis.upenn.edu

Abstract   In this paper we present a distributed multi-robot system that can achieve collaborative manipulation tasks through coordinated sensing, communication and control. We examine the specific task of locating an object in an unknown environment, caging it and then transporting it to a specified goal. Our framework for multi-robot coordination is extended from our previous work on cooperative localization and formation control. We focus on implementation and integration of these capabilities to robustly achieve the desired manipulation task.

Keywords: Multi-robot, formation control, cooperative sensing, manipulation

1.   INTRODUCTION

There are several tasks that can be performed more efficiently and robustly using multiple robots (Parker, 2000). We are motivated by tasks where robots can cooperate to manipulate and transport objects without using special purpose effectors or material handling accessories. In grasping and manipulation tasks, form and force closure properties and grasp stability lead to important constraints in manipulation (Howard and Kumar, 1996; Trinkle, 1992). There is also a significant body of literature in which the quasi-static assumption is used effectively to develop a paradigm for multi-robot manipulation (Mataric et al., 1995; Rus et al., 1995). In this paradigm, the robots can cooperatively push an object, generally maintaining a specified orientation to a goal position. In such situations, it is necessary to monitor the position and orientation of the manipulated object and to ensure that the perturbations caused by pushing are relatively small so that dynamics can be safely ignored.

In contrast, we propose a paradigm in which the manipulated object can be trapped or caged by a group of robots in formation, and the control of the flow of the group allows the transportation or manipulation of the grasped object. In this paradigm, the dynamics of the object and the robot-object interactions

A.C. Schultz and L.E. Parker (eds.), Multi-Robot Systems: From Swarms to Intelligent Automata, 93-100.
© 2002 *Kluwer Academic Publishers. Printed in the Netherlands.*

are never modeled, as is the case, for example in (Sugar and Kumar, 2000). Instead, by guaranteeing the shape of the formation, we can keep the manipulated object trapped amidst the robots. In contrast to other approaches to caging (Rimon and Burdick, 1998), we do not require conditions for form closure to be maintained. Neither do we need to plan the manipulation task as in (Sudsang and Ponce, 2000). Given the shape of the object, we can use well-known algorithms for planning inescapable grasps (Ponce and Faverjon, 1995) to derive the shape of the formation, and the allowable tolerance on shape changes.

In this paper we describe a methodology for combining cooperative sensing and formation control strategies that permits a group of nonholonomic mobile robots to achieve a coordinated manipulation task. We consider a scenario with no global positioning system and use vision as our primary sensing modality. In the rest of the paper, we describe our hardware, our cooperative sensing and control algorithms, and finally our experimental implementation of a distributed manipulation task.

2.   BACKGROUND AND PREVIOUS WORK

2.1   Hardware platform

The GRASP Laboratory Clodbuster (CB) robots (Figure 1) served as the testbed for all experiments. The platform is based upon a commercially available radio controlled truck, with significant modifications.

Figure 1. The Clodbuster^{TM} platforms used for experiments.

The robots utilize integrated computing. P-III processors and PXC200 framegrabbers support vision and control algorithms, while communication is achieved via 802.11 wireless ethernet connectivity. Each platform incorporates an omni-directional camera as its sole sensor. The 360 degree field of view (FOV) makes this an ideal choice for cooperative sensing tasks as will be discussed further in Section 3.

2.2   Formation control

In (Desai et al., 1998) two basic leader-following control laws based on input-output feedback linearization are presented. Here, we briefly describe the *Sepa-ration Bearing Controller* (*SBC*) and the *Separation Separation Control-ler* (*SSC*) since we use these controllers for our multi-robot manipulation task.

Since the robots are nonholonomic car-like platforms with two independent inputs and because we use input-output linearization, we are able to regulate two outputs using each control law. The kinematics of the ith robot (denoted by R_i) are given by

$$\dot{x}_i = v_i \cos \theta_i, \quad \dot{y}_i = v_i \sin \theta_i, \quad \dot{\theta}_i = \omega_i, \tag{1}$$

where $x_i \equiv (x_i, \ y_i, \ \theta_i) \in SE(2)$, and v_i and ω_i are the linear and angu-lar velocities, respectively. Using the *Separation Bearing Controller* (denoted

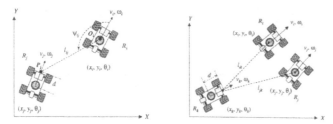

Figure 2.   Leader–following configurations, $SB_{ij}C$ (left) and $SS_{ijk}C$ (right)

$SB_{ij}C$ here), robot R_j can follow R_i with a desired *Separation* l_{ij}^d and desired relative *Bearing* ψ_{ij}^d. Note that this relative bearing describes the heading direc-tion of the follower with respect to the leader. Similarly, using the *Separation Separation Controller* (denoted $SS_{ijk}C$ here), robot R_k can follow both R_i (with desired *Separation* l_{ik}^d) and R_j (with desired *Separation* l_{jk}^d). In both cases, the closed-loop linearized system can be shown to be stable under suit-able assumptions on the leader's motion (Fierro et al., 2001).

Maintaining a rigid formation.   Using the controllers discussed in the previous subsection, the robots can maintain a prescribed rigid formation. In Figure 3, the initial team configuration is centered around the box, with the goal to flow the now encumbered formation along a trajectory generated by the leader. The motion of the leader with the formation constraints results in the object being dragged to a goal position. Note that the manipulation behavior is clearly open-loop while the formation control is a closed loop behavior.

Changing to a desired formation.   An example of adaptation of the formation is shown in Figure 4. Initially (top left), the robots are in a triangular formation with robot 2 (to the right of lead robot) employing a separation-bearing (SB_{12}) controller, and robot 3 (left) using a separation-separation (SS_{123}) controller. The detection of obstacles in front triggers a change in control assignment, when robot 3 switches its controller to a separation-bearing (SB_{13}) controller to enable a shape change from a triangular shape to a straight line. In this case, the determination of the target shapes (straight-line and triangle) is done off-line.

Figure 3.   Distributed manipulation demonstration with a team of three robots. Several snapshots from a sample run are shown here.

Figure 4.   Experiment demonstrating a change from a triangular formation to an in-line formation to avoid obstacles.

3.   COOPERATIVE SENSING MODES

Localization.   Using their omnidirectional camera systems, the robots are able to measure relative azimuth and elevation angles to each other. With the configuration space restricted to $SE(2)$, these are sufficient to estimate the relative positions of teammates. Orientations are still unknown. However, by exchanging sensor measurements any pair of robots can also estimate relative orientations. We term this procedure *pair-wise localization.* By chaining together pairs of these mutually visible robots, we recover the relative pose for the entire team. Our localization scheme extends our previous work (Spletzer and Taylor, 2000; Spletzer et al., 2001) by providing an efficient, scalable means for estimating the pose (and when not possible, the position) of a team of robots directly from the available sensor measurements.

We view the team as a directed *visibility graph* $G(V, E)$ in which the robots themselves serve as the vertices V. Edges E are defined dynamically based upon robot visibility. An edge $e_{ij} \in E$ if robot v_j is visible from robot v_i. Visibility is not commutative due to differences in sensor range, possible self-occlusions, and corrupted image data.

Next we employ a breadth-first search (BFS) based approach to explore the graph. Visited nodes are only enqueued if the corresponding robots are mutually visible. This can be determined in constant time by maintaining the graph in adjacency-matrix form. If mutual visibility is verified, the node is enqueued and the corresponding robot pose estimated. If the pair is not mutually visible *and* the visited node was not previously discovered, it is marked as such and the *position* of the corresponding robot estimated. This allows the node to be enqueued later in the search in the event that its orientation can also be estimated via a different path. This procedure requires $V - 1$ relative localizations and at most an additional $V - 2$ position estimates. As with BFS, it ensures that each node is enqueued no more than once. Additionally, the number of cycle checks cannot exceed the total number of edges E, and can be done in constant time. As a result, the procedure runs in $O(V + E)$ time.

Target Tracking. If the physical geometry of a target is known, a single robot can estimate its relative position. This constraint cannot be assumed for generic, unknown targets in the environment. For these, we view the robot team as a multi-eyed stereo rig with dynamic baselines obtained through the localization process. We then employ a least-squares estimator which utilizes all available azimuth angle measurements to maximize the quality of the target's position estimate.

When tracking an object with multiple point features, we assume that the azimuth angles to these points can be estimated from image features. The problem of tracking an object can be then be posed as tracking multiple point targets. We rely on work by Sogo *et al* to resolve the stereo correspondence problem (Sogo et al., 2000), where stereo pairs are validated with a third camera. In this way, multiple targets can be correctly associated by identifying the set of correspondences that minimizes the azimuth angle errors in the third image.

4. COOPERATIVE MANIPULATION

The configuration space of a planar object is a subset of $SE(2)$, the Special Euclidean group in the plane. The robots can be viewed as obstacles that constrain the object. It is meaningful to define the free space of an object as the set of all points in $SE(2)$ that are free of intersections of the object with one or more robots.

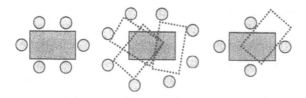

Figure 5. Examples of desirable (left, center) and undesirable (right) multi-robot formations for manipulation.

As Figure 5 illustrates there are three possible situations we are interested in. On the extreme left is a formation that corresponds to a point in a compact subset of the free space. The object is completely trapped by the surrounding robots. The middle formation is similar, except the object can rotate and occupy any orientation. On the extreme right is a formation in which the object can escape the surrounding robots if it is allowed to rotate. Obviously, if the object can move relative to the objects and the dynamics of the object is unknown, this formation is not desirable. Our goal is to achieve formations like the ones shown in Figure 5 (left, center) and control the team of robots in the prescribed formation to move them to a desired destination.

Figure 6.   Experiment demonstrating a 4 robot team locating a box, capturing it and transporting it.

Experiments.   A cooperative manipulation task can be decomposed into four phases with associated switches in behavior. These are: object discovery, approach, trapping, and manipulation. The four phases are shown in Figure 6 for an experiment with four robots manipulating a box, and where one robot is initially in close proximity to the box. In the discovery phase (top left), the robot team searches for the target object while maintaining an in-line formation. Upon box discovery, a formation switch is triggered and the task enters the approach phase (top right). The goal of this phase is to generate robot trajectories to desired positions around the target to facilitate caging. The controllers must rely on both relative localization estimates of the formation and cooperative estimates of the target position.

Once the robots reach their desired positions and orientations with respect to the target, it is trapped within the formation (bottom left). With the trapping phase completed, the robot team marches with the object along a leader-generated trajectory without having to servo on the box position (bottom left). Regulating the formation shape is sufficient to ensure that the object remains trapped. Snapshots of cooperative localization data from the same experiment are shown in Figure 7.

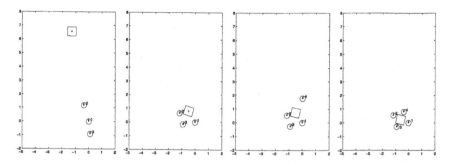

Figure 7. Experimental localizer data for the four mode switches. All positions and orientations are shown relative to robot 2 (origin for the plots)

5. CONCLUSIONS

In this paper we presented a distributed multi-robot system that can successfully execute cooperative manipulation tasks that involve – locating an object in an unknown environment, trapping it and transporting it to a specified goal. Our approach is based on two sets of previous results: formation control (Desai et al., 1998; Fierro et al., 2001) and cooperative localization (Spletzer and Taylor, 2000; Spletzer et al., 2001). We are pursuing experiments with larger teams of robots while developing algorithms for discovery and leader election that will allow ad-hoc networking and self-organization of the team.

Acknowledgments

This work was supported by the DARPA ITO MARS Program, grant 130-1303-4-534328-xxxx-2000-0000, NSF grant CDS-97-03220 and AFOSR grant F49620-01-1-0382. We thank Rafael Fierro, Jim Ostrowski, Peng Song and Zhidong Wang for discussions on formation control and distributed manipulation.

References

Desai, J., Ostrowski, J. P., and Kumar, V. (1998). Controlling formations of multiple mobile robots. In *Proc. IEEE Int. Conf. Robot. Automat.*, pages 2864–2869, Leuven, Belgium.

Fierro, R., Das, A., Kumar, V., and Ostrowski, J. P. (2001). Hybrid control of formation of robots. To appear in IEEE Int. Conf. Robot. Automat.

Howard, W. S. and Kumar, V. (1996). On the stability of grasped objects. *IEEE Trans. Robot. Automat.*, 12(6):904–917.

Mataric, M., Nilsson, M., and Simsarian, K. (1995). Cooperative multi-robot box pushing. In *IEEE/RSJ International Conf. on Intelligent Robots and Systems*, pages 556–561, Pittsburgh, PA.

Parker, L. E. (2000). Current state of the art in distributed autonomous mobile robotics. In Parker, L. E., Bekey, G., and Barhen, J., editors, *Distributed Autonomous Robotic Systems*, volume 4, pages 3–12. Springer, Tokyo.

Ponce, J. and Faverjon, B. (1995). On computing three finger force-closure grasp of polygonal objects. *IEEE Trans. Robot. Automat.*, 11(6):868–881.

Rimon, E. and Burdick, J. W. (1998). Mobility of bodies in contact–i: A new 2nd order mobility index for multiple-finger grasps. *IEEE Trans. Robot. Automat.*, 2(4):541–558.

Rus, D., Donald, B., and Jennings, J. (1995). Moving furniture with teams of autonomous robots. In *IEEE/RSJ International Conf. on Intelligent Robots and Systems*, pages 235–242, Pittsburgh, PA.

Sogo, T., Ishiguro, H., and Trivedi, M. (2000). Real-time target localization and tracking by n-ocular stereo. In *IEEE International Conf. on Computer Vision*.

Spletzer, J., Das, A. K., Fierro, R., Taylor, C. J., Kumar, V., and Ostrowski, J. (2001). Cooperative localization and control for multi-robot manipulation. In *IEEE/RSJ International Conf. on Intelligent Robots and Systems*.

Spletzer, J. and Taylor, C. (2000). Cooperative localization using omnidirectional vision. Technical Report MS-CIS-01-12, GRASP Laboratory, University of Pennsylvania, PA. Available at http://www.cis.upenn.edu/techreports.html.

Sudsang, A. and Ponce, J. (2000). A new approach to motion planning for disc-shaped robots manipulating a polygonal object in the plane. In *Proc. IEEE Int. Conf. Robot. Automat.*, pages 1068–1075, San Francisco, CA.

Sugar, T. and Kumar, V. (2000). Control and coordination of multiple mobile robots in manipulation and material handling tasks. In Corke, P. and Trevelyan, J., editors, *Experimental Robotics VI: Lecture Notes in Control and Information Sciences*, volume 250, pages 15–24. Springer-Verlag.

Trinkle, J. C. (1992). On the stability and instantaneous velocity of grasped frictionless objects. *IEEE Trans. Robot. Automat.*, 8(5).

IV

COORDINATION AND FORMATIONS

A LAYERED ARCHITECTURE FOR COORDINATION OF MOBILE ROBOTS

Reid Simmons, Trey Smith, M. Bernardine Dias,
Dani Goldberg, David Hershberger, Anthony Stentz, Robert Zlot
Robotics Institute, Carnegie Mellon University
Pittsburgh, PA 15213
{ reids,trey,mbdias,danig,hersh,axs,robz } @ri.cmu.edu

Abstract   This paper presents an architecture that enables multiple robots to explicitly coordinate actions at multiple levels of abstraction. In particular, we are developing
an extension to the traditional three-layered robot architecture that enables robots
to interact directly at each layer – at the behavioral level, the robots create distributed control loops; at the executive level, they synchronize task execution;
at the planning level, they use market-based techniques to assign tasks, form
teams, and allocate resources. We illustrate these ideas through applications in
multi-robot assembly, multi-robot deployment, and multi-robot mapping.

Keywords: Multi-robot coordination, robot architecture, task-level control.

1.   INTRODUCTION

An architecture for multi-robot coordination must be able to accommodate
issues of synchronization and cooperation under a wide range of conditions
and at various levels of granularity and timescales. At the concrete, physical
level of sensors and actuators, the robots need to respond quickly to dynamic
events (such as imminent collisions), while at the same time reasoning about
and executing long-term strategies for achieving goals. Executing such strategies is likely to involve establishing and managing a variety of synchronization constraints between robots. For some tasks, such as distributed search,
coordination may be loose and asynchronous; for others, such as cooperative
manipulation of a heavy object, coordination must be tightly synchronized.

In designing a multi-robot architecture that allows flexibility in synchronization and granularity, one must inevitably negotiate the tension between
centralized and distributed approaches. A centralized system can make optimal decisions about subtle issues involving many robots and many tasks. In

A.C. Schultz and L.E. Parker (eds.), Multi-Robot Systems: From Swarms to Intelligent Automata, 103-112.
© 2002 *Kluwer Academic Publishers. Printed in the Netherlands.*

contrast, a highly distributed system can quickly respond to problems involving one, or a few, robots and is more robust to point failures.

We are developing a multi-robot coordination architecture that addresses these issues, providing flexibility in granularity and synchronization, while also attempting to accommodate the strengths of both distributed and centralized approaches. The architecture is an extension of the traditional three-layered approach, which provides event handling at different levels of abstraction through the use of behavioral, executive, and planning layers. Our approach extends the architecture to multiple robots by allowing robots to interact directly at each layer (see Figure 1). This provides several benefits, including (1) plans can be constructed and shared between multiple robots, using a market-based approach, providing for various degrees of optimization; (2) executive-level, inter-robot synchronization constraints can be established and maintained explicitly; and (3) distributed behavior-level feedback loops can be established to provide for both loosely- and closely-coupled coordination.

Each layer is implemented using representations and algorithms that are tuned to the granularity, speed, and types of interactions typically encountered at each level (symbolic/global, hybrid/reactive, numeric/reflexive). The strengths of this architecture are its flexibility in establishing interactions between robots at different levels and its ability to handle tasks of varying degrees of complexity while maintaining reactivity to changes and uncertainty in the environment. This paper presents the major components of the architecture, together with case studies that illustrate their use in multi-robot applications.

2.   RELATED WORK

Our approach blends the advantages of both the centralized and distributed approaches to multi-robot systems. In the centralized approach, a centralized planner plans out detailed actions for each robot. For example, a planner might treat two 6 DOF arms as a single 12 DOF system for the purpose of generating detailed trajectories that enable the arms to work together in moving some object, without bumping into each other (Khatib, 1995). While this approach provides for close coordination, it does so at the expense of local robot autonomy. In particular, this approach usually employs centralized monitoring and, if anything goes wrong, the planner is invoked to replan everything. Thus, this approach suffers from single point failure and lack of local reactivity.

At the other end of the spectrum, in the distributed approach (Arkin, 1992; Balch and Arkin, 1994; Mataric, 1992; Parker, 1998) each agent is autonomous, but there is usually no explicit synchronization among the robots. Coordination (or, more accurately, cooperation) occurs fortuitously, depending on how the behaviors of the robots interact with the environment. For instance, in the

ALLIANCE architecture (Parker, 1998), robots decide which tasks to perform in a behavior-based fashion: They have "motivations" that rise and fall as they notice that tasks are available or not. While ALLIANCE can handle heterogeneous robots (robots can have different motivations for different tasks), it does not deal with the problem of explicit coordination.

(Jennings and Kirkwood-Watts, 1998) have developed a distributed executive for multi-robot coordination. The executive, based on a distributed dialect of Scheme, is similar to our executive language in the types of synchronization constructs it supports. As with our work, this enables robots to solve local coordination problems without having to invoke a high-level planner.

Several researchers have investigated economy-based architectures applied to multi-agents systems (Sandholm and Lesser, 1995; Sycara and Zeng, 1996; Wellman and Wurman, 1998), beginning with work on the Contract Net (Smith, 1980). (Golfarelli et al., 1997) proposed a negotiation protocol for multi-robot coordination that restricted negotiations to task-swaps. (Stentz and Dias, 1999) proposed a more capable market-based approach that aims to opportunistically introduce pockets of centralized planning into a distributed system, thereby exploiting the desirable properties of both distributed and centralized approaches. (Thayer et al., 2000; Gerkey and Matarić, 2001; Zlot et al., 2002) have since presented market-based multi-robot coordination results.

3. APPROACH

Our multi-robot architecture is based on the layered approach that has been adopted for many single-agent autonomous systems (Bonasso et al., 1997; Muscettola et al., 1998; Simmons et al., 1997). These architectures typically consist of a planning layer that decides how to achieve high-level goals, an executive layer that sequences tasks and monitors task execution, and a behavioral layer that interfaces to the robot's sensors and effectors.

Typically, information and control flows up and down between layers. The planning layer sends plans to the executive, which further decomposes tasks into subtasks and dispatches them based on the temporal constraints imposed by the plan. Dispatching a task often involves enabling or disabling various behaviors. The behaviors interact to control the robot, sending back sensor data and status information. The executive informs the planner when tasks are completed, and may further abstract sensor data for use by the planner. The executive also monitors task execution: In case of failure, it can try to recover or it can terminate the task and request a new plan from the planner.

We extend this architectural concept to multiple robots in a relatively straightforward way. Each robot is composed of a complete three-layered architecture. In addition, each of the three layers can interact directly with the same layer of the other robots (Figure 1). Thus, each robot can act autonomously at all

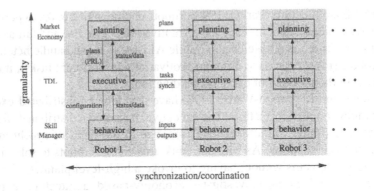

Figure 1.   Layered multi-robot architecture

times, but can coordinate (at multiple levels) with other agents, when needed. By allowing each layer to interact directly with its peers, we can essentially form distributed feedback loops, operating at different levels of abstraction and at different timescales. In particular, the behavioral layer coordinates behaviors, the executive layer coordinates tasks, and the planning layer coordinates/schedules resources. In this way, problems arising can be dealt with at the appropriate level, without having to involve higher layers. This decreases latency and may increase robustness (since lower layers typically operate with higher-fidelity models).

The following sections describe each of the layers of the architecture in more detail. For each, we also provide a case study illustrating how interaction at that level can be used for multi-robot coordination.

3.1   Coordinated Behaviors

The behavioral layer consists of real-time sensor/effector feedback loops. By connecting the sensor behaviors of one robot to the effector behaviors of another, we can create efficient distributed servo loops (Simmons et al., 2000b). Similarly, by connecting effector behaviors together, we can create tightly coordinated controllers. For instance, two robots could coordinate their arm and navigation behaviors to jointly carry a heavy piece of equipment (Pirjanian et al., 2001).

A multi-robot behavioral layer that can support such capabilities needs several critical features. First, it must be possible to connect behaviors to one another, enabling sensor data and status information to flow between behaviors on different robots. The architecture should not place restrictions on the type of data that can pass between distributed behaviors. Also, it should be possible to connect behaviors transparently on different robots, in the same way that one connects them on the same robot. Finally, high-bandwidth, low-latency communications is needed to achieve good performance in interacting, multi-robot behaviors.

Our implementation extends the Skill Manager of (Bonasso et al., 1997) to provide for both intra- and inter-robot connections. Skills are connected via input/output ports and operate in a data-flow fashion: When a new input value arrives on a port, the skill runs an action code that (optionally) produces outputs. For skills on the same robot, the connection is via function call; for inter-robot connections, data flows using a transparent message-passing protocol. Certain aspects of the skills, such as the action code and number and types of ports, are statically determined at compile time. Most aspects, however, can be dynamically configured at run time (either from the executive layer, or from a skill's action code). These include the ability to enable and disable a behavior, set the value of an input, set parameters of the skill, and create or destroy the connections between ports. The executive layer can also subscribe to skill outputs.

These ideas have been demonstrated in the context of distributed visual servoing for large-scale assembly using multiple, heterogeneous robots (Simmons et al., 2000b). The task, which is to move the end of a large beam into a vertical stanchion, uses three robots (Figure 2) – an observer (a mobile robot with stereo vision) and two controllers (an automated crane and a mobile manipulator).

The observer robot tries to maintain the best view of the fiducials on the beam, stanchion, and manipulator arm. It move the cameras and drives around the workspace to keep those fiducials it is currently tracking centered in the image and filling most of the cameras' fields of view. The observer uses stereo vision to compute the 6 DOF pose of the fiducials and outputs the differences between the poses to one of the controller robots, depending on which skill it is connected to at the time. In particular, the observer's visual tracking behavior (Figure 3) first helps direct the crane to move the beam near the stanchion, then aids the mobile manipulator in grabbing the beam, and finally helps the mobile manipulator to place the beam in the stanchion (5mm clearance).

3.2   Coordinated Task Execution

The executive layer has responsibility for hierarchically decomposing tasks into subtasks, enforcing synchronization constraints between tasks (both those imposed by the planner and those added during task decomposition), monitoring task execution, and recovering from exceptions (Simmons, 1994).

For coordinated multi-robot task execution, it should be possible to synchronize tasks transparently on two different agents, in the same way as if they were performed by a single agent. For instance, we may want to state that a robot should not start analyzing a rock until two other rovers have moved into place

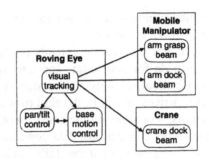

Figure 2.   Heterogeneous robots for   *Figure 3.*   Coordinated behaviors for
large-scale assembly   distributed visual servoing

to provide assistance. In addition, a distributed executive should facilitate one robot monitoring the execution of another robot and helping it recover from faults. Finally, it is desirable for one robot to be able to directly request that another robot perform a task (such as assisting it to perform visual servoing).

Our interface between planner and executive is based on PRL, a plan representation language that was developed for an earlier project (Simmons et al., 2000a). PRL represents plans as a hierarchy of tasks, with each task defined in terms of parameters, subtasks, and temporal constraints between the subtasks. We have recently extended PRL to enable specification of the resource utilization of a task and which agent should be executing it.

The executive is implemented using the Task Description Language. TDL is an extension of C++ that contains explicit syntax to support hierarchical task decomposition, task synchronization, execution monitoring, and exception handling (Simmons and Apfelbaum, 1998). Recently, we have extended TDL to handle task-level coordination between robots and to enable one robot to spawn or terminate a task on another. TDL transparently handles passing task data (function parameters) and synchronization signals between robots, using message passing.

We have used these ideas to perform coordinated deployment of multiple, heterogeneous robots (Simmons et al., 2000a). The executive receives a plan consisting of a set of deployments, where each deployment is an ordered list of locations, the robots that should to deploy to those locations, and a deployment "style." For instance, in the "group" deployment style all the robots navigate concurrently to the first deployment location, one stays behind while the others continue to the next location, and so on. The executive has procedures for decomposing each deployment style into primitive navigation behaviors. It is then responsible for coordinating the tasks, to ensure that robots do not move until others are in position. Figure 4 presents a simplified version of the TDL code for deploying the robots shown in Figure 5.

```
Goal GroupDeploy (DEPLOY_PTR deployList) {
  with (serial) {
    for (int i=0; i<length(deployList); i++) {
      spawn GroupDeploySub(i, deployList)
} } }

Goal GroupDeploySub (int phase,
                     DEPLOY_PTR deployList) {
  with (parallel) {
    for (int j=phase; j<length(deployList); j++) {
      spawn Navigate(deployList[i].location)
                     with on deployList[j].robot;
} } }
```

Figure 4.   TDL code (simplified) for "group" deployment strategy

Figure 5.   Coordinated deployment of heterogeneous robots

3.3   Coordinated Planning

Our approach to task allocation and planning is based on a market economy. An *economy* is essentially a population of agents coordinating with each other to produce an aggregate set of goods. *Market economies* are those generally unencumbered by centralized planning, instead leaving individuals free to exchange goods and services and enter into contracts as they see fit. Despite the fact that individuals in the economy act only to advance their own self-interests, the aggregate effect is a highly productive society.

We have developed a market-based architecture in which tasks are allocated based on exchanges of single tasks between pairs of robots. A robot that needs a task performed announces that it will auction off the task as a buyer. Each seller capable of performing the task for a cost c, bids to do so for $c + \epsilon$. The buyer accepts the lowest bid, as long as it is cheaper than doing the task itself. If a bid is accepted, the seller performs the task and pockets ϵ as profit.

This approach has been demonstrated in several simulated and actual robot applications. (Zlot et al., 2002) reports on multi-robot exploration of an unknown environment. Each robot generates a list of target points to visit, and orders them into a tour (using an approximate TSP algorithm). Next, the robots try to auction off tasks, one at a time (sequentially along the tour). When all its auctions close, the robot navigates to its first target point, incorporates map information, and generates new target points. Note that the robots are able to share map information via trades on the market, and that all communication is asynchronous and not assumed to be reliable. Figure 6 shows a map built by four robots in a highbay, and Figure 7 shows the paths they took.

More recently, our market-based architecture has been extended to support *leader* agents that engage in multi-agent, multi-task exchanges. This can lead to task allocations that outperform those produced by repeatedly exchanging one task at a time. A leader agent opens a *combinatorial exchange* in which agents can bid to buy or sell combinations of tasks. The leader chooses which

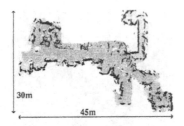

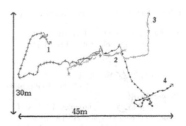

Figure 6. Map of highbay created by four robots (from (Zlot et al., 2002))

Figure 7. Paths taken by exploring robots (from (Zlot et al., 2002))

bids to accept and pockets the difference between the total revenue and cost of all the transactions. In order to earn a good profit, however, it must perform the computationally difficult problem of choosing which bids to accept in order to maximize efficiency. A leader opportunistically announces exchanges for those pockets of the overall problem that will net a good return on its computational investment (those that can produce the largest efficiency gain from its optimization).

3.4   Capability Server

A mechanism for providing up-to-date information on robots and their capabilities is necessary for highly dynamic multi-robot systems, where robot attrition, accession, and modification are common. The Agent Capability Server (ACS) is designed to handle this by providing a distributed facility for automatically disseminating agent information. The idea is that each robot in the system has its own local ACS that it can quickly query for the capabilities and status of other robots. This information can then be used in determining which robots can perform which tasks, planning efficient multi-robot strategies for a task, establishing the appropriate executive-level and behavior-level coordination mechanisms between robots, and accommodating non-responsive (possibly failed) robots. Our ACS is similar in purpose, though specialized by comparison, to middle agents that map capabilities to particular agents, and agent naming services that map agents to locations (Sycara et al., 2001).

Maintaining the consistency of information among the various Agent Capability Servers is a key concern, especially since communications range may be limited and influenced by geographic features. Thus, it cannot be assumed that a broadcast of new information will reach all robots. One method for boosting consistency in the face of uncertain communication is to have each ACS broadcast its data periodically. Thus, new information will eventually propagate throughout the group by transparently using robots as communication relays.

Each ACS monitors the periodic updates from the other servers. When a server has failed to provide an update for a sufficiently long interval, the robot is assumed to have failed or be completely out of communications range. In either case, its entry is removed from the ACS. If and when the missing robot is heard from again, it will be seamlessly re-incorporated.

4. SUMMARY AND FUTURE WORK

This paper describes a multi-robot extension to the traditional three-layered architecture, where each layer can communicate directly with its peer layers on other robots. This gives the robots the ability to coordinate at multiple levels of abstraction with minimal overhead in terms of inter- and intra-agent communication. We have described criteria and design decisions for each layer, and have presented case studies showing how layer-to-layer interaction enables reliable multi-robot coordination.

Much work still remains on the architecture. We need to integrate the planning and executive layers more fully. We need to generalize the market-based planning framework, especially by adding leader agents and their more sophisticated bidding structures. We need to have all levels of the architecture deal more fully with the loss of agents. We need to complete design and implementation of the Agent Capability Server. And, we need to demonstrate the architecture in rich domains.

Multi-robot coordination promises huge benefits in terms of increased capability and reliability. The price, however, is often added complexity. We believe that a well-structured, flexible architecture will facilitate the development of such systems, at reasonable cost.

Acknowledgments

This work has been supported by several grants, including NASA NCC2-1243, NASA NAG9-1226, DARPA N66001-99-1-892, and DARPA DAAE07-98-C-L032. Thanks go to Steve Smith, Jeff Schneider, Drew Bagnell, and Vincent Cicirello for their comments and advice on the architecture. David Apfelbaum implemented and helped design TDL.

References

Arkin, R. (1992). Cooperation without communication: Multiagent schema-based robot navigation. *Journal of Robotic Systems*, 9(3):351–364.

Balch, T. and Arkin, R. C. (1994). Communication in reactive multiagent robotic systems. *Autonomous Robots*, 1(1):27–52.

Bonasso, R., Kortenkamp, D., Miller, D., and Slack, M. (1997). Experiences with an architecture for intelligent, reactive agents. *Journal of Artificial Intelligence Research*, 9(1).

Gerkey, B. P. and Matarić, M. J. (2001). Sold!: Market methods for multi-robot control. In *IEEE Transactions on Robotics and Automation Special Issue on Multi-Robot Systems*.

Golfarelli, M., Maio, D., and Rizzi, S. (1997). A task-swap negotiation protocol based on the contract net paradigm. Technical Report CSITE, 005-97, University of Bologna.

Jennings, J. and Kirkwood-Watts, C. (1998). Distributed mobile robotics by the method of dynamic teams. In *Proc. Conference on Distributed Autonomous Robot Systems*.

Khatib, O. (1995). Force strategies for cooperative tasks in multiple mobile manipulation systems. In *Proc. International Symposium of Robotics Research*.

Mataric, M. (1992). Distributed approaches to behavior control. In *Proc. SPIE Sensor Fusion V*, pages 373–382.

Muscettola, N., Nayak, P. P., Pell, B., and Williams, B. (1998). Remote agent: To boldly go where no ai system has gone before. *Artificial Intelligence*, 103(1–2):5–48.

Parker, L. (1998). ALLIANCE: An architecture for fault tolerant multirobot cooperation. *IEEE Transactions on Robotics and Automation*, 14(2):220–240.

Pirjanian, P., Huntsberger, T., and Barrett, A. (2001). Representation and execution of plan sequences for distributed multi-agent systems. In *Proc. International Conference on Intelligent Robots and Systems*.

Sandholm, T. and Lesser, V. (1995). Issues in automated negotiation and electronic commerce: Extending the contract net framework. In *Proc. International Conference on Multiagent Systems*, pages 328–335.

Simmons, R. (1994). Structured control for autonomous robots. *IEEE Transactions on Robotics and Automation*, 10(1):34–43.

Simmons, R. and Apfelbaum, D. (1998). A task description language for robot control. In *Proc. International Conference on Intelligent Robots and Systems*, Vancouver Canada.

Simmons, R., Apfelbaum, D., Fox, D., Goldman, R., Haigh, K. Z., Musliner, D., Pelican, M., and Thrun, S. (2000a). Coordinated deployment of multiple, heterogeneous robots. In *Proc. International Conference on Intelligent Robots and Systems*, Takamatsu Japan.

Simmons, R., Goodwin, R., Haigh, K., Koenig, S., and O'Sullivan, J. (1997). A layered architecture for office delivery robots. In *Proc. First International Conference on Autonomous Agents*.

Simmons, R., Singh, S., Hershberger, D., Ramos, J., and Smith, T. (2000b). First results in the coordination of heterogeneous robots for large-scale assembly. In *Proc. International Symposium on Experimental Robotics*, Honolulu Hawaii.

Smith, R. (1980). The contract net protocol: High-level communication and control in a distributed problem solver. *IEEE Transactions on Computers*, C-29(12):1104–1113.

Stentz, A. and Dias, M. B. (1999). A free market architecture for coordinating multiple robots. Technical Report CMU-RI-TR-99-42, Robotics Institute, Carnegie Mellon University.

Sycara, K., Paolucci, M., van Velsen, M., and Giampapa, J. (2001). The retsina mas infrastructure. Technical Report CMU-RI-TR-01-05, Robotics Institute, Carnegie Mellon University.

Sycara, K. and Zeng, D. (1996). Coordination of multiple intelligent software agents. *International Journal of Cooperative Information Systems*, 5(2–3).

Thayer, S., Digney, B., Dias, M. B., Stentz, A., Nabbe, B., and Hebert, M. (2000). Distributed robotic mapping of extreme environments. In *Proceedings of SPIE: Mobile Robots XV and Telemanipulator and Telepresence Technologies VII*.

Wellman, M. and Wurman, P. (1998). Market-aware agents for a multiagent world. *Robotics and Autonomous Systems*, pages 115–125.

Zlot, R., Stentz, A., Dias, M. B., and Thayer, S. (2002). Multi-robot exploration controlled by a market economy. In *Proc. International Conference on Robotics and Automation*.

STABILITY ANALYSIS OF DECENTRALIZED COOPERATIVE CONTROLS

John T. Feddema, David A. Schoenwald
Sandia National Laboratories[2]
Intelligent Systems and Robotics Center
Albuquerque, NM 87185
{Jtfedde, daschoe}@sandia.gov

Abstract: This paper describes how large-scale decentralized control theory may be used to analyze the stability of multiple cooperative robotic vehicles. Models of cooperation are discussed from a decentralized control system point of view. Whereas decentralized control research in the past has concentrated on using decentralized controllers to partition complex physically interconnected systems, this work uses decentralized methods to connect otherwise independent non-touching robotic vehicles so that they behave in a stable, coordinated fashion. A vector Liapunov method is used to prove stability of two examples: the controlled motion of multiple vehicles along a line and the controlled motion of multiple vehicles in a plane.

Keywords: Mobile robotics, optimal cooperative controls

1.   INTRODUCTION

Most recently, researchers have begun to investigate using decentralized control techniques to control multiple vehicles. Chen and Luh (Chen and Luh, 1994) examined decentralized control laws that drove a set of mobile robots into a circle formation. Similarly, Yamaguchi studied line-formations

[2] Sandia is a multi-program laboratory operated by Sandia Corporation, a Lockheed Martin Company, for the United States Department of Energy under contract DE-AC04-94AL85000. This research is partially funded by the Information Technology Office of the Defense Advanced Research Projects Agency.

A.C. Schultz and L.E. Parker (eds.), Multi-Robot Systems: From Swarms to Intelligent Automata, 113-122.
© 2002 *Kluwer Academic Publishers. Printed in the Netherlands.*

(Yamaguchi and Arai, 1994) and general formations (Yamaguchi and Burdick, 1998), and so did (Yoshida, *et al.*, 1994). Decentralized control laws using a potential field approach to guide vehicles away from obstacles can be found in (Molnar and Starke, 2000; Schneider, *et al.*, 2000). Beni and Liang (Beni and Liang, 1996) prove the convergence of a linear swarm of distributed autonomous vehicles into a synchronously achievable configuration.

In this paper, we address the stable control of multiple vehicles using large-scale decentralized control techniques (Siljak, 1991). In a previous paper, we described how to test for controllability and observability of a large-scale system (Feddema and Schoenwald, 2001). Once we know that a system is structurally observable and controllable, the next question to ask is that of connective stability. Will the overall system be globally asymptotically stable under structural perturbations? Analysis of connective stability is based upon the concept of vector Liapunov functions, which associates several scalar functions with a dynamic system in such a way that each function guarantees stability in different portions of the state space. The objective is to prove that there exist Liapunov functions for each of the individual subsystems and then prove that the vector sum of these Liapunov functions is a Liapunov function for the entire system.

2.   STABILITY OF LARGE SCALE SYSTEMS

Suppose that the overall system is denoted by
$$\mathbf{S}: \quad \dot{x} = f(t,x,u)$$
$$y = h(t,x) \tag{1}$$

where $x(t) \in \Re^n$ is the state of $\mathbf{S}$ (e.g., x, y position, orientation, and linear and angular velocities of all vehicles) at time $t \in T$, $u(t) \in \Re^m$ are the inputs (e.g., the commanded wheel velocities of all vehicles), and $y(t) \in \Re^\ell$ are the outputs (e.g., GPS measured x,y position of all vehicles). The function $f: T \times \Re^n \times \Re^m \to \Re^n$ describes the dynamics of $\mathbf{S}$, and the function $h: T \times \Re^n \to \Re^\ell$ describes the observations of $\mathbf{S}$. We can partition the system into N interconnected subsystems given by
$$\mathbf{S}: \quad \dot{x}_i = f_i(t,x_i,u_i) + \widetilde{f}_i(t,x,u), \qquad i \in \{1,...,N\}$$
$$y_i = h_i(t,x_i) + \widetilde{h}_i(t,x) \tag{2}$$

where $x_i(t) \in \Re^{n_i}$ is the state of the ith subsystem $\mathbf{S}_i$ at time $t \in \Re$, $u_i(t) \in \Re^{m_i}$ are the inputs to $\mathbf{S}_i$, and $y_i(t) \in \Re^{\ell_i}$ are the outputs of $\mathbf{S}_i$. The function $f_i : T \times \Re^{n_i} \times \Re^{m_i} \to \Re^{n_i}$ describes the dynamics of $\mathbf{S}_i$, and the function $\widetilde{f}_i : T \times \Re^n \times \Re^m \to \Re^{n_i}$ represents the dynamic interaction of $\mathbf{S}_i$ with the rest of the system $\mathbf{S}$. The function $h_i : T \times \Re^{n_i} \to \Re^{\ell_i}$ represents observations at $\mathbf{S}_i$ derived only from local state variables of $\mathbf{S}_i$, and the function $\widetilde{h}_i : T \times \Re^n \to \Re^{\ell_i}$ represents observation at $\mathbf{S}_i$ derived from the rest of $\mathbf{S}$. The N independent subsystems are denoted as

$$\mathbf{S}_i : \quad \dot{x}_i = f_i(t, x_i, u_i), \qquad i \in \{1, \ldots, N\}$$
$$y_i = h_i(t, x_i). \tag{3}$$

Both local and interconnected feedback may be added to the system with

$$u_i = k_i(t, y_i) + \widetilde{k}_i(t, y), \qquad i \in \{1, \ldots, N\} \tag{4}$$

where the function $k_i : T \times \Re^{\ell_i} \to \Re^{m_i}$ represents the feedback at $\mathbf{S}_i$ derived only from local observations, and the function $\widetilde{k}_i : T \times \Re^\ell \to \Re^{m_i}$ represents the feedback at $\mathbf{S}_i$ derived from the rest of $\mathbf{S}$. For stability analysis, we will assume that the control function has already been chosen and the closed loop dynamics of the system can be written as

$$\mathbf{S} : \quad \dot{x}_i = g_i(t, x_i) + \widetilde{g}_i(t, x), \qquad i \in \{1, \ldots, N\}. \tag{5}$$

where the function $g_i : T \times \Re^{n_i} \to \Re^{n_i}$ describes the closed loop dynamics of $\mathbf{S}_i$. The closed loop interconnection function can be written as

$$\widetilde{g}_i(t, x) = \widetilde{g}_i(t, \overline{e}_{i1} x_1, \overline{e}_{i2} x_2, \ldots, \overline{e}_{iN} x_N), \qquad i \in \{1, \ldots, N\} \tag{6}$$

where $\overline{e}_{ij} \in B^{n_i \times n_j}$, and the elements of the fundamental interconnection matrix $\overline{E} = (\overline{e}_{ij})$ are

$$(\overline{e}_{ij})_{pq} = \begin{cases} 1, & (x_j)_q \text{ occurs in } (\widetilde{g}_i(t, x, u))_p \\ 0, & (x_j)_q \text{ does not occur in } (\widetilde{g}_i(t, x, u))_p. \end{cases} \tag{7}$$

where $q \in \{n_j\}$ and $p \in \{n_i\}$.

The structural perturbations of $\mathbf{S}$ are introduced by assuming that the elements of the fundamental interconnection matrix that are one can be replaced by any number between zero and one, i.e.

$$e_{ij} = \begin{cases} [0,1], & \overline{e}_{ij} = 1 \\ 0, & \overline{e}_{ij} = 0. \end{cases} \tag{8}$$

Therefore, the elements e_{ij} represent the strength of coupling between the individual subsystems. A system is connectively stable if it is stable in the sense of Liapunov for all possible $E = \left(e_{ij}\right)$ (Siljak, 1991). In other words, if a system is connectively stable, it is stable even if an interconnection becomes decoupled, i.e. $e_{ij} = 0$, or if interconnection parameters are perturbed, i.e. $0 < e_{ij} < 1$. This is potentially very powerful, as it proves that the system will be stable if an interconnection is lost through communication failure.

For a possibly nonlinear system **S** to be connectively stable, there must exist a matrix $W = \left(w_{ij}\right)$ that is an M-matrix (i.e. all leading principal minors must be positive):

$$w_{ij} = \begin{cases} 1 - \bar{e}_{ii}\kappa_i\xi_{ii}, & i = j \\ -\bar{e}_{ij}\kappa_i\xi_{ij}, & i \neq j \end{cases} \tag{9}$$

where $\kappa_i > 0$, and the scalar function $v_i : T \times \mathfrak{R}^{n_i} \to \mathfrak{R}_+$ must satisfy a Lipschitz condition

$$\left|v_i\left(t,x'\right) - v_i\left(t,x_i''\right)\right| \leq \kappa_i \left\|x_i' - x_i''\right\|, \quad \forall t \in T, \quad \forall x_i', x_i'' \in \mathfrak{R}^{n_i}. \tag{10}$$

Also, the constant $\xi_{ij} \geq 0$ for $i \neq j$ and satisfy

$$\left\|\tilde{g}_i\left(t,x\right)\right\| \leq \sum_{j=1}^{N} \bar{e}_{ij}\xi_{ij}\phi_j\left(\left\|x_j\right\|\right), \quad \forall(t,x) \in T \times \mathfrak{R}^n \tag{11}$$

where the time derivative of the Liaponuv function is less than the negative of the comparison function $\phi_j\left(\left\|x_j\right\|\right)$

$$\dot{v}_i\left(t,x_i\right) \leq -\phi_j\left(\left\|x_j\right\|\right), \quad \forall(t,x_i) \in T \times \mathfrak{R}^{n_i}. \tag{12}$$

For linear systems, the matrix W is a function of the eigenvalues of the state transition matrix. Suppose the linear system dynamics are

$$\mathbf{S}: \quad \dot{x}_i = A_i x_i + \sum_{j=1}^{N} e_{ij} A_{ij} x_j, \quad i \in \{1,...,N\}, \tag{13}$$

and the Liapunov function for each individual subsystem is $v_i\left(x_i\right) = \left(x_i^T H_i x_i\right)^{1/2}$ where H_i is a positive definite matrix. For the system **S** to be connectively stable, the following test matrix $W = \left(w_{ij}\right)$ must be an M-matrix (Siljak, 1991):

$$w_{ij} = \begin{cases} \dfrac{\lambda_m\left(G_i\right)}{2\lambda_M\left(H_i\right)} - \bar{e}_{ii}\lambda_M^{1/2}\left(A_{ii}^T A_{ii}\right), & i = j \\ -\bar{e}_{ij}\lambda_M^{1/2}\left(A_{ij}^T A_{ij}\right), & i \neq j \end{cases} \tag{14}$$

where the symmetric positive definite matrix G_i satisfies the Liapunov matrix equation $A_i^T H_i + H_i A_i = -G_i$, and $\lambda_m(\bullet)$ and $\lambda_M(\bullet)$ are the minimum and maximum eigenvalues of the corresponding matrices. This same analysis can also be performed in the discrete domain (Sezer and Siljak, 1988).

2.1   Example of a Linear Interconnected System

As an example, let us analyze a simple linear one-dimensional problem in which a chain of interdependent vehicles is to spread out along a line as shown in Figure 1(a). The objective is to spread out evenly along the line using only information from the nearest neighbor.

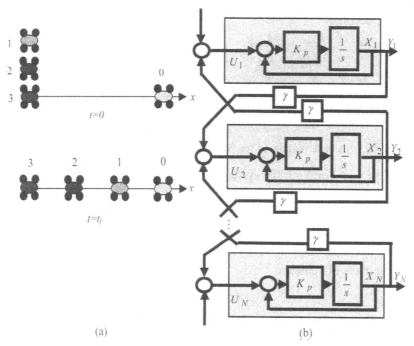

(a)   (b)

Figure 1. (a) One-dimensional control problem. The top line is the initial state. The second line is the desired final state. Vehicles 0 and 3 are boundary conditions. Vehicles 1 and 2 spread out along the line by using only the position of their left and right neighbor. (b) Control block diagram of N-vehicle interaction problem.

Assume that the vehicle's plant is modeled as a simple integrator, and the commanded input is the desired velocity of the vehicle along the line. A feedback loop and a proportional gain K_p are used to control each vehicle's position (see Figure 1(b)). The dynamics of each subsystem is

$$S_i: \quad \dot{x}_i = -K_p x_i + K_p u_i, \qquad i \in \{1,...,N\}$$

$$y_i = x_i \tag{15}$$

where x_i is the position of the ith vehicle, u_i is the control input, and y_i is the observation. Assume the control of each vehicle is a function of the two nearest vehicles' observed positions, and the boundary conditions on the first and last vehicle are 1 and 0, respectively.

$$u_1 = 1 + \gamma y_2$$

$$u_i = \gamma(y_{i-1} + y_{i+1}) \qquad i \in \{2,...,N-1\} \tag{16}$$

$$u_N = \gamma y_{N-1}$$

where γ is the interaction gain between vehicles. For this linear system, the test matrix becomes

$$W = \begin{bmatrix} K_p & -K_p\gamma & 0 & \cdots & & 0 \\ -K_p\gamma & K_p & -K_p\gamma & & & \vdots \\ 0 & -K_p\gamma & K_p & & & 0 \\ \vdots & & & \ddots & & -K_p\gamma \\ 0 & \cdots & & 0 & -K_p\gamma & K_p \end{bmatrix}. \tag{17}$$

For $N=2$, this test matrix is an M-matrix (i.e. the system is connectively stable) if $|\gamma| < 1$. For $N=3$, the system is connectively stable if $|\gamma| < \frac{1}{\sqrt{2}}$. For $N=4$, the system is connectively stable if $|\gamma| < 0.618$. Notice how the range of the interaction gain gets smaller for larger sized systems. In fact, for this particular example, the interaction gain range reaches a limit of $|\gamma| \le 0.5$ for infinite numbers of vehicles.

2.2   Example of a Non-Linear Interconnected System

Next, let us consider the problem of N vehicles spreading out in a two-dimensional space while staying a specified distance from their neighbors (See Figure 2). We assume that the vehicles communicate their position to their neighbors and that each vehicle knows the distance that it is suppose to be from neighboring vehicles. Is there a decentralized control that will drive the group of vehicles to the desired configuration?

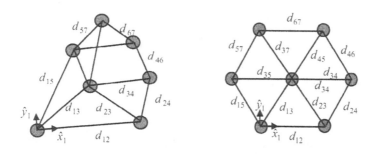

Figure 2. (a) Initial configuration of vehicles. (b) Desired configuration.

To solve such a problem, a gradient-based control law is proposed and a vector Liapunov technique (Siljak, 1991) is used to prove stability. The dynamics of the vehicles are essentially ignored so that the vehicle dynamics can be considered to be the control law only.

$$\dot{\overline{x}}_i = \overline{u}_i, \quad i = 1,\ldots,N \tag{18}$$

where $\overline{x}_i = [x_i \, y_i]^T \in \mathfrak{R}^2$ is the *i*th vehicle state space vector, and $\overline{u} \in \mathfrak{R}^2$ is the control input. The scalar values x_i and y_i are the *x* and *y* coordinates of the *i*th vehicle. A Liapunov function is defined for each vehicle that is minimized when the vehicle is a specified distance from the other vehicles.

$$v_i(t,\overline{x}_i) = \sum_{j=1}^{N} e_{ij} \left[d_{ij}^2 - (x_i - x_j)^2 - (y_i - y_j)^2 \right]^2 \tag{19}$$

where

$$e_{ij} = \begin{cases} 0, & j \text{ is not nearest neighbor} \\ 1, & j \text{ is nearest neighbor} \end{cases} \tag{20}$$

and d_{ij} are the desired distances between the *i* and *j* vehicles (note $d_{ij} = d_{ji}$ and $d_{ii} = 0$). The decentralized Liapunov functions v_i are a measure of the sum of the squared errors in distance for vehicle *i* with respect to all the neighboring vehicles. Since this function is not zero at $\overline{x}_i = 0$, a new state vector $\tilde{x}_i = \overline{x}_i - \overline{x}_{i0}$ is defined such that

$$d_{ij}^2 = \left\| \overline{x}_{i0} - \overline{x}_{j0} \right\|^2 \tag{21}$$

where $\overline{x}_{i0}$ is the final position of the *i*th vehicle after the vehicles are dispersed and is considered a constant. Then the Liapunov function for the *i*th vehicle can be written as

$$v_i(t,\tilde{x}_i) = \sum_{j=1}^{N} e_{ij}\left[-\left\|\tilde{x}_i - \tilde{x}_j\right\|^2 - 2(\tilde{x}_i - \tilde{x}_j)^T(\bar{x}_{i0} - \bar{x}_{j0})\right]^2 \qquad (22)$$

which equals zero when $\tilde{x}_i = 0$ and is greater than zero for $\tilde{x}_i \neq 0$.

In order to minimize the ith vehicle Liapunov function, we use a control law that is the spatial gradient of the Liapunov function:

$$\dot{\tilde{x}}_i = \bar{u}_i = -\alpha \frac{\partial v_i(t,\tilde{x}_i)}{\partial \tilde{x}_i}, i = 1,...,N \qquad (23)$$

where $\alpha > 0$ is the control gain. The time derivative of the ith vehicle Liapunov function is given by

$$\dot{v}_i(t,\tilde{x}_i) = \frac{\partial v_i(t,\tilde{x}_i)}{\partial t} + \frac{\partial v_i(t,\tilde{x}_i)}{\partial \tilde{x}_i}\dot{\tilde{x}}_i = -\alpha\left\|\frac{\partial v_i(t,\tilde{x}_i)}{\partial \tilde{x}_i}\right\|^2 \qquad (24)$$

where $\qquad (25)$

$$\frac{\partial v_i(t,\tilde{x}_i)}{\partial \tilde{x}_i} = -4\sum_{j=1}^{N} e_{ij}\left(d_{ij}^2 - \left\|x_i - \tilde{x}_j\right\|^2\right)(\tilde{x}_i - \tilde{x}_j)$$

$$= -4\sum_{j=1}^{N} e_{ij}\left[-\left\|\tilde{x}_i - \tilde{x}_j\right\|^2 - 2(\tilde{x}_i - \tilde{x}_j)^T(\bar{x}_{i0} - \bar{x}_{j0})\right]\left[(\tilde{x}_i - \tilde{x}_j) + (\bar{x}_{i0} - \bar{x}_{j0})\right]$$

Since $\dot{v}_i \leq 0$ and it is equal to zero only at $\tilde{x}_i = 0$, this is a valid Liapunov function and the gradient-based control law is stable for a single vehicle.

The next step is to show that when all vehicles use the same control law that the entire system is stable. We assume that the Liapunov function for the entire system can be described as a vector Liapunov function (the sum of the individual Liapunov functions)

$$v(t,\tilde{x}) = \sum_{j=1}^{N} \rho_j v_j(t,\tilde{x}_j) \qquad (26)$$

where $\rho_j > 0$. Clearly, $v \geq 0$ for all $\tilde{x} \in \Re^n$ and it is equal to zero only if $\tilde{x} = 0$. We want to show that $\dot{v}(t,\tilde{x}) \leq 0$ for all $\tilde{x} \in \Re^n$ and it is equal to zero only if $\tilde{x} = 0$. The time derivative of the vector Liapunov function is

$$\dot{v}(t,\tilde{x}) = \frac{\partial v(t,\tilde{x})}{\partial t} + \left[\frac{\partial v(t,\tilde{x})}{\partial \tilde{x}}\right]^T \dot{\tilde{x}} \qquad (27)$$

Since $v(t,\tilde{x})$ is independent of time, the first term on the right is zero. If $\rho_i = 1$ and $e_{ij} = e_{ji}$ for all $i, j \in N$, then the second term is

$$\dot{v}(t,\tilde{x}) = -16\alpha\sum_{i=1}^{N}\eta_i \qquad (28)$$

where

$$\eta_i = z_i^T X_i^T X_i z_i \tag{29}$$

$$X_i = [(\bar{x}_i - \bar{x}_1) \quad \cdots \quad (\bar{x}_i - \bar{x}_N)] \in \Re^{2 \times N} \tag{30}$$

$$z_i = \left[e_{i1}\left(d_{i1}^2 - \|\bar{x}_i - \bar{x}_1\|^2\right) \quad \cdots \quad e_{iN}\left(d_{iN}^2 - \|\bar{x}_i - \bar{x}_N\|^2\right) \right]^T \in \Re^{N \times 1} \tag{31}$$

Since $X_i^T X_i$ is a positive semi-definite matrix, then $\eta_i \geq 0$ for $i \in \{1,...,N\}$ and $\dot{v}(t,\tilde{x}) \leq 0$ for all $\tilde{x} \in \Re^n$. The elements of this semi-definite matrix are

$$\left(X_i^T X_i\right)_{pq} = \|\bar{x}_i - \bar{x}_p\| \|\bar{x}_i - \bar{x}_q\| \cos\theta_{piq} \tag{32}$$

where θ_{piq} is the angle between vectors $\bar{x}_i - \bar{x}_p$ and $\bar{x}_i - \bar{x}_p$. The derivative of the vector Liapunov is equal to zero only when $\tilde{x} = 0$, which is the same as $d_{ij}^2 - \|\bar{x}_i - \bar{x}_j\|^2 = 0$ or $z_i = 0$ for all $i,j \in \{1,...,N\}$. This proves that the system Liapunov function $v(t,\tilde{x})$ is valid, and the decentralized gradient-based control law drives the entire system to a stable configuration.

3.   CONCLUSIONS

In this paper, we mathematically described how to determine if a cooperative robotic system is connectively stable. We illustrated the use of this technique on both a linear and a non-linear problem. The control law for the linear problem has been applied to robotic perimeter surveillance task. The control law for the non-linear problem has been applied to a building surveillance task. Hardware implementation of these control algorithms will be presented at the conference.

References

Beni, G. and Liang, P. (1996). Pattern Reconfiguration in Swarms – Convergence of a Distributed Asynchronous and Bounded Iterative Algorithm. In IEEE Transactions on Robotics and Automation, 12(3):485-490.

Chen, Q. and Luh, J.Y.S. (1994). Coordination and Control of a Group of Small Mobile Robots. In *Proceedings of IEEE International Conference on Robotics and Automation*, 3:2315-2320.

Feddema, J. T., and Schoenwald, D. A. (2001). Decentralized Control of Cooperative Robotic Systems. In *Proceedings of SPIE*, 4364, Orlando, FL.

Molnar, P. and Starke, J. (2000). Communication Fault Tolerance in Distributed Robotic Systems. *Distributed Autonomous Robotic Systems 4*, Parker, L. E., Bekey, G., and Barhen, J., editors, Springer-Verlag, pages 99-108.

Schneider, F. E., Wildermuth, D., Wolf, H.-L. (2000). Motion Coordination in Formations of Multiple Robots Using a Potential Field Approach. *Distributed Autonomous Robotic Systems 4*, Parker, L. E., Bekey, G., and Barhen, J., editors, Springer-Verlag, pages 305-314.

Sezer, M. E. and Siljak, D. D. (1988). Robust Stability of Discrete Systems. *International Journal of Control*, 48(5):2055-2063.

Siljak, D. D. (1991). Decentralized Control of Complex Systems. Academic Press.

Yamaguchi, H. and Arai, T. (1994). Distributed and Autonomous Control Method for Generating Shape of Multiple Mobile Robot Group. In *Proceedings of the IEEE International Conference on Intelligent Robots and Systems*, 2:800-807.

Yamaguchi, H., and Burdick, J.W. (1998). Asymptotic Stabilization of Multiple Nonholonomic Mobile Robots Forming Group Formations. In *Proceedings of the 1998 Conference on Robotics & Automation*, pages 3573-3580, Leuven, Belgium.

Yoshida, E., Arai, T., Ota, J., and Miki, T. (1994). Effect of Grouping in Local Communication System of Multiple Mobile Robots. In *Proceedings of the IEEE International Conference on Intelligent Robots and Systems*, 2:808-815.

SNOW WHITE AND THE 700 DWARVES
A Cooperating Robotic System

Brian H. Wilcox

Supervisor, Robotic Vehicles Group,
Jet Propulsion Laboratory,
California Institute of Technology

Abstract: Detailed scientific study of possible sites on Mars such as extinct hot springs may require robotic excavation of the sort generally associated with archaeological "digs" on Earth. This paper describes a system developed to explore the feasibility of building such a robotic network. A scaling analysis shows that, for a given landed mass, it is better to break the payload into as many small vehicles as possible, due to the surface-to-volume implications on the power-to-weight ratio for the system using solar power. Each small vehicle must be equipped with a shovel and hopper to move loose material, and a provision for a percussive hammer to break up hard soils and rocks. A central mast rising from the nearby lander supports stereo imaging and a computer providing central co-ordination of activities. Each vehicle has limited sensing (e.g. no vision) but is designed to follow simple commands (goto(x,y)) and rules (keep to the right of oncoming traffic). The outer surface of each vehicle is arranged to capture solar energy, and the lander mast can be configured with a foil reflector to bounce sunlight down into the excavation to keep the vehicles powered most of the day. Thus we have a system consisting of a single tall mast with a pair of stereo cameras at the top and a large, silver reflector hanging down, together with a large number of small robots, each having a pick and shovel. This system has been dubbed "Snow White and the 700 Dwarves".

Keywords: Co-operating Robots, excavation robots

A.C. Schultz and L.E. Parker (eds.), Multi-Robot Systems: From Swarms to Intelligent Automata, 123-130.

1. INTRODUCTION

1.1 Development of the Concept

Previous work from the NASA Nanorover Technology task (Wilcox, *et al.*, 1997) and the MUSES-CN flight project (Wilcox and Jones, 2000) has developed the key technologies for a nanorover qualified for space flight (~1kg, Figure 1) which combines advanced mobility (4-wheels with active articulation, self-righting), advanced sensing (capacitive proximity sensing on each wheel and accurate laser ranging to 10 meters), advanced cryovac microactuators (-180C to +125C operational thermal range), and advanced chip-on-board, wide-temperature electronic packaging of a 32-bit flight computer with a custom flight gate array I/O. This foundation can be used to develop a first example of a robotic outpost system that could perform a useful function on Mars or other planetary surface.

Figure 1. Nanorover Prototype, approximately 20 cm long overall

The mission concept for this system is to have a solar concentrator deployed from a mast on the lander. This reflector concentrates sunlight onto a small region of the ground. A group of solar powered rovers gather at this site and use percussive devices to excavate terrain material. Each has solar panels on the exterior of a "dump truck" vehicle configuration, where the excavated terrain is stored for transport. Each small rover might have a mass of approximately 1 kg, while 1000 kg or more may be landed on Mars by a single lander, so it may be possible to put hundreds of these small vehicles on

Mars in a single mission. This multitude of small rovers with percussive rock breaking devices, together with the solar concentrator tower, has been dubbed "Snow White and the 700 Dwarves".

The objective of the present activity is to explore the scaling relations and a distributed control methodology needed to achieve this objective. Specifically, there are many scaling advantages for using large numbers of small solar powered robots, most importantly the increased power to weight ratio and amortization of development costs over a large number of vehicles. In this task, four testbed vehicles have been built to perform an initial demonstration of excavating an open pit (Figure 2).

Figure 2. Nanorover Outpost Vehicles in various phases of operation.

1.2   Mission Applications for Co-operating Robots

Excavation of open trenches is of profound interest to both the Planetary Science and Human Exploration and Development of Space (HEDS) communities due to it's applicability in study of areas of intense science interest, such as extinct hydrothermal deposits, as well as emplacement of habitats under radiation shielding, digging utility trenches, or extraction and processing of the near-surface volatiles thought to exist at some latitudes on Mars. Excavation makes an ideal first co-operative task for a robotic outpost: it scales extremely well to small sized vehicles, it requires co-operation to be effective, and it can be done with solar power and thus eliminates the need for an elaborate means for storing and distributing energy to the multitude of vehicles. Many other proposed robot outpost applications do not exhibit any "advantage of scale" in that they merely multiply the capabilities of a single rover proportionately, such as in parallel exploration of a site by many uncoordinated explorers. The present concept is most attractive in conjunction with a solar concentrator tower, allowing a relatively inexpensive distribution of power to the vehicles. This combination, "Snow White and the 700 Dwarves" allows many exciting missions on Mars, in the polar regions on Earth's moon, at the polar regions on Mercury, and perhaps on comets and asteroids. There is an especially exciting possibility with respect to the lunar polar regions where, because of the small angle between the moon's equator and the ecliptic plane, a tower only a few hundred meters high could intercept continuous sunlight which could be reflected 24-hours a day down into foil-lined tunnels excavated into the putative permafrost. Such a tower would be relatively easy to implement in the low gravity, disturbance-free environment of the moon.

The fleet of dwarf rovers could excavate these tunnels first for science, looking at a layered history of cometary impacts on the moon and possibly even examine history of the larger asteroidal impact events on Earth. (It is possible that the impact event thought to have caused the extinction of the dinosaurs may have lofted sufficient seawater to have left a record of volatiles and possibly even microfossils deposited in the cold traps at the lunar poles.) Moreover, these tunnels could provide vast quantities of water and perhaps other volatiles for use in life support, material processing, and propulsion for future HEDS endeavors. If the quantities of volatiles prove adequate, these lunar polar bases could well become the springboard for human settlement of the entire solar system, due to the resulting low cost to send large quantities of reaction mass and life support fluids to interplanetary or Earth orbits. These activities would in turn be enabled by the hard-working fleet of dwarf robots tirelessly toiling in the reflected light of Snow White.

1.3   Scaling Laws for Planetary Rovers

It is clear that the surface area, and therefore the solar power collection area, of a rover scales with the square of its dimension, assuming fixed proportions. The mass, weight in a gravity field, and resistance to motion scale with the cube of the dimension. Therefore the power to weight ratio scales inversely with the dimension, so that small rovers have relatively large power-to-weight ratios and visa versa. This power-to-weight ratio is the main determinant of the overall vehicle performance, just as with terrestrial vehicles. For example, given a fixed coefficient of rolling resistance, the maximum speed of a vehicle is linear with the power-to-weight ratio. This leads us to the somewhat counterintuitive result that, for solar powered vehicles, a small vehicle can go twice as fast (even up a hill) as one twice as big. Or it can carry twice the load, in proportion to its size. This further leads us to the conclusion that, if we can only land a fixed mass on the surface of Mars, we would be well advised to break that mass into as many small vehicles as possible.

Of course, there is a limit to this. As a practical matter, there is overhead mass associated with the control system, the power system, and the structure which is not necessarily scale invariant. Small motors tend to be less efficient than large ones, small batteries have a greater fraction of their mass devoted to inert packaging, etc. This means that, given the current state of technology, there is an optimal size for the vehicle.

More importantly, the forces needed to excavate rock or packed soil do not scale well to small vehicles. The use of percussion breaks this adverse quasi-static scaling relationship, allowing the advantages of small robots to be applied to the task of excavation. Small vehicles have a difficult time generating sufficient forces to excavate any material, even in the smaller unit quantities accepted by the small body. Percussion offers a way out of this dilemma. With percussion, the peak forces that can be exerted by the vehicle are no longer limited to some small multiplier of the vehicle's weight (which is itself reduced in planetary environments over that on Earth). In a percussive system, this multiple is increased by the ratio of the hammer forward stroke distance to its stopping distance. For hard rock or packed regolith (planetary dust/sand or other particulates) this ratio can be very large, of order 1000 or more. This means that a hammer can momentarily impart forces of 1000 times the weight of the vehicle to this type of terrain. If the surface area of the hammer face is small, this means that pressures far in excess of the compressive strength of the terrain can be exerted, which pulverizes the medium into a network of microcracks.

2.   APPROACH

The basic principles to be explored by this task are that small, distributed, solar powered robots with percussive excavation tools can perform soil/rock excavation and manipulation tasks better than larger or quasi-static systems for a given launch mass. The approach is to create a small fleet of miniature robotic vehicles incorporating the essential features described here: a low-power "dump truck" with a hammer for breaking the terrain medium, a scoop for loading it into the hopper, and appropriate minimalist sensing and computing environment to allow an effective collective behavior in excavating a single, coherent "dig". This small fleet is used to develop the needed software to allow a demonstration of the enormous scaling advantages of the distributed excavation approach.

2.1   Vehicle and System Description

The vehicles themselves are four-wheeled "rocker bogie" vehicles (Bickler, 1992) manufactured using 3-D photolithography for the body and wheels, with machined metal parts for the running gear and scoop assemblies. The electronics consists of a commercial equivalent of the rad-hard R3000 32-bit processor developed by Synova running under the VxWorks real-time operating system on a custom printed wiring board with all the relevant robotic I/O. Each individual rover has pitch and roll sensors, motor position and current sensing, and provision for capacitive proximity sensing in each wheel. Each vehicle has a 2-DOF scoop able to swing entirely around the vehicle, enabling excavation of loose material, dumping of that material, self-righting of the vehicle from any orientation. Also, a central system "Snow White" has been built which allows co-ordination of the vehicles under command from a human operator. A custom-developed radio system allows communication between Snow White and each of the rovers on a single channel. This radio, built using a chip set from RF Monolithics, Inc., gives 9600 baud communications at ranges of about 100 meters.

The control methodology for this system is strictly hierarchical. Snow White is responsible for maintaining an estimate of the position of each vehicle, and commanding them to their next task. Co-operative excavation occurs because Snow White commands successive rovers to dig in slightly different spots that, taken together, take on the geometry of the finished excavation. Each vehicle only is responsible for dead reckoning a relatively short distance, since it has no sensors to correct for wheel slip or other navigation errors. Snow White, on the other hand, has a global sensor (the mast mounted cameras) in a fixed co-ordinate system, and so can maintain a coherent long-term representation of the state of the entire system.

The software functionality implemented to date on each vehicle includes operator-designated waypoint dead reckoning (using differential odometry for heading estimation). Scooping, dumping, self-righting, and motor stall recovery behaviors have been implemented and tested. As with Sojourner, which melded the behavior-control approach with "waypoint designation" to allow an effective command-level interface, this system will continue to evolve that methodology with "rules of the road" (e.g. "pass to the right") (Wilcox and Nguyen, 1997). A custom communication protocol has been developed which allows one or more rovers to act as intermediate relay nodes to pass messages to any rover which strays too far from Snow White.

Software that has been implemented for Snow White includes "blob tracking" of one or multiple rovers by the Snow White cameras. This algorithm relies on the fact that the only parts of the scene that move are the rovers. Each rover is tracked to ensure that it is staying on the commanded sequence of waypoints. Other software that has been developed and tested includes provision for simultaneous teleoperation of multiple rovers, update of individual rover calibration parameters, patching of the flash memory, and rover engineering downlink telemetry interpretation, display, and logging.

3. CONCLUSIONS AND FUTURE WORK

The hierarchical organization of this system is a natural consequence of the nature of long-distance communications. Because such communications (e.g. between Earth and Mars) are necessarily energy intensive and require significant infrastructure, there is generally only a single asset at any site which is capable of providing such a service. With only one element of the system able to receive human commands, it is natural for that element to be the local "commander". In this case, it is the lander that delivers the rovers to the surface and out of which pops the mast for Snow White. It seems quite possible that this same "communications bottleneck" applies to many sorts of systems, such as military applications. In that event, it seems possible that this hierarchical structure would be appropriate where the commander has the long-range radio and the "foot soldiers" have only local communications. While it is true that such a hierarchical organization might be less fault tolerant than one where any element can take over any job function, in fact the uniqueness of the long range communication asset implicitly makes the system asymmetric and inhomogeneous. The advantage of such a hierarchical system is that it is clear how to develop the software to make it work.

In the next year of funding, it is expected that behaviors will be implemented incorporating the capacitive proximity sensing functions built into the existing hardware. This hardware excites each wheel as a "free space" capacitor, and can detect very small changes in the capacitance of each wheel. This allows estimation of the proximity and amount of dielectric material in proximity to the wheel. For example, if a wheel starts to go off the edge of a cliff, the capacitance will drop abruptly even before the wheel loses contact with the terrain. The capacitance will increase if the wheel is against an obstacle, or near to another vehicle. This sensor data will be used to implement behaviors including the "pass to the right" function mentioned earlier as well as cliff avoidance.

Also, the percussive mechanisms for breaking up rocks and "hard pan" soils will be added to the vehicles. The dynamics of reaction of the vehicle to both the forward stroke of the hammer and to the resulting impact have not been explored. One intriguing approach is to have a hammer on the end of a swing arm, which can also then be used as a scooping device. The hammer can be accelerated upwards on one end of the semicircular stroke, so that the reaction force on the rover body is downwards, preventing unwanted rover motion. The hammer can then coast around the arc to either a vertical face impact or further around to hammer on the horizontal ground, or anywhere in between. Minimal reaction forces from the hammer impact will be imparted back to the rover via the relatively long, free swing arm.

In conclusion, the key figure of merit for the overall system is the mass of excavated material per unit system mass (rovers plus additional infrastructure) per unit time. The activity will be judged successful if the scaling advantages of the fleet of small rovers are realized: that the fleet of rovers are able to excavate many times their own mass per year of operation.

Acknowledgments

The research described in this publication was carried out by the Jet Propulsion Laboratory, California Institute of Technology, under contract with the National Aeronautics and Space Administration. The author (the principal investigator for this effort) gratefully acknowledges the contributions of Greg Levanas (cognizant engineer), Tom McCarthy (systems engineer), Jack Morrison (software engineer), John Lo and Steve Kondos (mechanical engineers) and Hung Tran (technician) without whose conscientious and tireless efforts this activity would not have been possible.

References

Bickler, D. (1992). The new family of JPL planetary vehicles. In *Proceedings of International Symposium on Planetary Mobile Vehicles,* Toulouse-Labege France.

Wilcox, B. H., Nasif, A. K., and Welch, R. V. (1997). A Nanorover for Mars. *Space Technology,* 17 (3-4):163-172.

Wilcox, B. H., and Nguyen, T. (1997). Sojourner on Mars and Lessons Learned for Future Planetary Rovers. *Society of Automotive Engineers publication* 981695.

Wilcox, B. H., and Jones, R. M. (2000). The MUSES-CN Nanorover Mission and Related Technology. In Siegwart, R., editor, *Proceedings IEEE International Conference on Robotics and Automation, Workshop on Mobile Micro-Robots,* San Francisco, CA.

V

SENSOR AND HARDWARE ISSUES

GOATS: MULTI-PLATFORM SONAR CONCEPT FOR COASTAL MINE COUNTERMEASURES

Henrik Schmidt and Joseph R. Edwards

Massachusetts Institute of Technology

Cambridge, MA 02139 USA

henrik@keel.mit.edu, jre@mit.edu

Abstract   Recent progress in underwater robotics and acoustic communication has led to the development of a new paradigm in ocean science and technology, the Autonomous Ocean Sampling Network (AOSN). AOSN consists of a network of fixed moorings and/or autonomous underwater vehicles (AUV), tied together by state-of-the-art acoustic communication technology. The GOATS'2000 (Generic Oceanographic Array Technology Systems) Joint Research Program is aimed toward the development of environmentally adaptive AOSN technology specifically directed toward Rapid Environmental Assessment and Mine Counter Measures in coastal environments. The research program combines theory and modeling of the 3-D environmental acoustics with experiments involving AOSN and sensor technology. [Work supported by ONR and SACLANT]

Keywords: Mine Countermeasures, Autonomous Underwater Vehicles, Rapid Environmental Assessment, Shallow water, Sonars, Oceanography.

1.   INTRODUCTION

Recent progress in underwater robotics and acoustic communication has led to the development of a new paradigm in ocean science and technology, the Autonomous Ocean Sampling Network (AOSN) (Curtain et al., 1993). AOSN consists of a network of fixed moorings and/or autonomous underwater vehicles (AUV) tied together by state-of-the-art acoustic communication technology.

The Generic Ocean Array Technology Sonar (GOATS) concept for coastal mine countermeasures (MCM) is a derivative of AOSN specifically aimed at detecting and classifying targets on and within the seabed in very shallow water (VSW). A fleet of AUV's connected by an underwater communication network and equipped with acoustic receiver arrays is used to measure the 3-D

A.C. Schultz and L.E. Parker (eds.), Multi-Robot Systems: From Swarms to Intelligent Automata, 133-140.
© 2002 *Kluwer Academic Publishers. Printed in the Netherlands.*

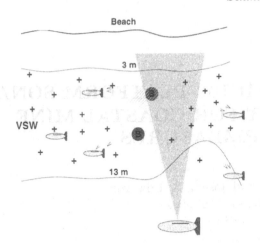

Figure 1.   GOATS: Generic Ocean Array Technology Sonar concept for coastal MCM. A fleet of AUV's connected by an underwater communication network, and equipped with acoustic receiver arrays is used to measure the 3-D scattering from proud and buried targets insonified by a dedicated master AUV.

scattering from proud and buried targets insonified by a low-frequency (1-20 kHz) projector mounted on a dedicated vehicle. The 3-D scattered field from seabed targets is target specific, and it is envisioned that by characterizing its spatial and temporal characteristics the fleet of AUV's may be capable of simultaneously detecting and classifying seabed objects. To optimally explore the acoustic signatures of the targets as classification, the bi-static sonar system should operate in the mid-frequency regime where both geometric and resonant target scattering are significant, which for meter size objects corresponds to the 1-20 kHz mid-frequency regime (Schmidt and Lee, 1999). This relatively low active sonar frequency regime is also highly beneficial in terms of bottom penetration (Maguer et al., 2000), suggesting that GOATS has significant potential for detection and classification of buried mines in very shallow water.

The potential of this new multi-static sonar concept for VSW MCM is explored as part of the GOATS'2000 Joint Research Project (JRP), a collaborative research effort between SACLANT Undersea Research Centre, MIT, WHOI, FAU, and several other institutions in the US and Europe (Schmidt and Bovio, 2000). The central element of the JRP is a prototype underwater vehicle network operated remotely from the SACLANTCEN oceanographic vessel R/V Alliance, including moored sources and arrays, AUVs equipped with a variety of sensors, and reliable navigation and communication systems. The major environmental acoustics objective of the JRP is to use an acoustic array mounted on an AUV to characterize the spatial and temporal characteristics of the 3-D scattering from seabed targets and the associated reverberation, including the effects of multipaths. This effort is aimed at establishing the foundation for future multi-static sonar concepts exploring 3-D acoustic signatures for combined detection and classification in very shallow water, as envisioned in Figure 1.

A major potential advantage of the AOSN concept is its adaptive sampling capabilities. The network can be designed to change its behavior dependent on the sensor responses. AUVs carrying MCM sonars can be programmed to change their survey patterns to optimize the classification of detected targets.

The GOATS effort will incrementally address the scientific and technological issues associated with the development of new AOSN-based MCM and REA concepts, including reliable communication and navigation capabilities. Building on the results of the GOATS'98 sea trial, the GOATS'2000 experiment is currently being carried out at Elba Island, Italy.

2. GOATS'98 EXPERIMENT

2.1 Odyssey II AUV

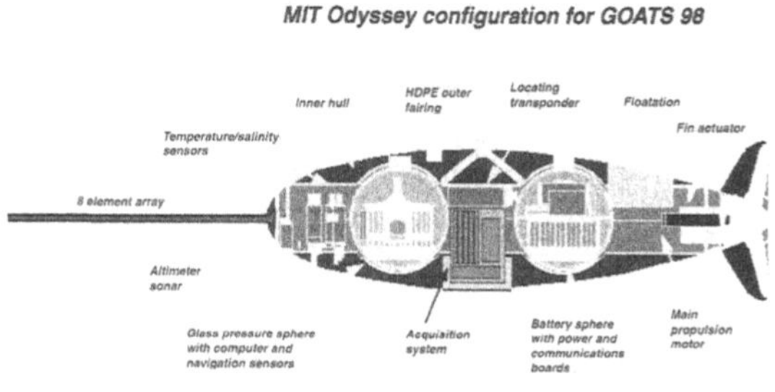

Figure 2. Configuration of Odyssey II AUV 'Xanthos' for acoustic measurements in GOATS'98. The AUV control electronics and batteries are located in two 17" Benthos glass spheres. An 8-element array is mounted in a 'swordfish' configuration, and connected to a dedicated acquisition system in the center well of the vehicle.

An Odyssey II class autonomous underwater vehicle was used as the platform for the mobile acoustic array in GOATS'98, Figure 2. A substantial fraction of the vehicle is dedicated to wet volume, which enables the Odyssey II to support a wide range of payload systems that include CTD, ADCP/DVL, ADV, side-scan sonar, USBL tracking systems, OBS, and several video systems. The core vehicle has a depth rating of 6,000 m, weighs 120 kg, and measures 2.2 m in length and 0.6 m in diameter. It cruises at approximately 1.5 m/s (3 knots) with endurance in the range of 3-12 hours, depending on the battery installed and the load. The AUV used in GOATS'98 featured an acoustic array for bistatic reception, developed at SACLANTCEN, consisting of a line array, mounted in the vehicle's nose, in a 'swordfish' configuration, and an autonomous data acquisition system, installed in a watertight canister in the vehicle's payload bay.

3. BISTATIC SYNTHETIC APERTURE SONAR PROCESSING

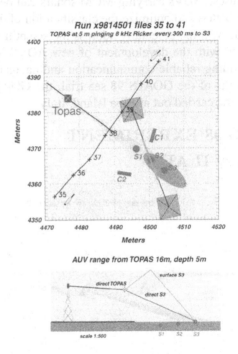

Figure 3.   Bistatic sonar geometry. The TOPAS parametric source is insonifying the seabed with a footprint of approximately 5 × 10, centered on the half-buried spherical target S3. The Spherical target S2 i flush buried.

Figure 3 shows the bistatic sonar geometry of Mission X14501 of the Goats'98 experiment. The TOPAS parametric source is insonifying the seabed with a footprint of approximately 5 × 10 , centered on the half-buried spherical target S3. The Spherical target S2 is flush buried. The Odyssey AUV is passing over the targets receiving the scattered field along its track, creating a synthetic aperture.

The synthetic aperture sonar (SAS) processing involved in the GOATS project differs from standard SAS imaging techniques in several important ways. The primary scientific issues that have been and will continue to be addressed are demonstration of the AUV as a viable and robust synthetic aperture sonar platform, extension of sonar processing techniques to bi- and multi-static scenarios and optimization of information processing for the robust detection and classification of man-made objects. To date, some progress has been made in all of these domains, although the final solutions remain a research concern.

The AUV has proven to be capable of providing a platform for synthetic aperture imaging (Edwards et al., 2001; LePage and Schmidt, 2001; Schmidt et al., 2000). Synthetic apertures of up to 10 times the physical aperture length have been used for imaging with the data received on the AUV-borne receiver. The maximum synthetic aperture length has in fact only been limited by the LBL navigation cycle that creates a gap in the acquired data. Such aperture extension provides both improved angular resolution and a significant reduction in the incoherent noise. Figure 4 shows a typical synthetic aperture image of the GOATS'98 target field, including the clear detection of a proud and a flush-buried sphere. The flush-buried sphere is detected despite the fact that it is insonified at a sub-critical grazing angle. Its location in the image is also slightly removed from its actual position. This apparent displacement was hypothesized to be caused by an unexpectedly strong delayed elastic return from the target. Frequency analysis and modeling results have since confirmed this theory.

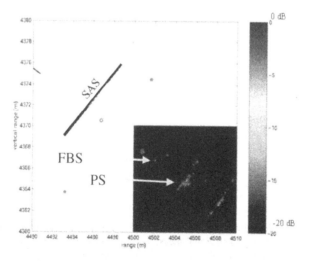

Figure 4. Bistatic synthetic aperture image of the GOATS'98 target field. A proud sphere (PS) and a flush-buried sphere (FBS) lie inside the source beam. Both are visible in the image, but the buried sphere appears displaced by about 1 m. This displacement is due to the fact that the imaged signal is a delayed elastic return, providing evidence for "anomalous" coupling between the evanescent sub-bottom field and the elastic target.

The extension of sonar processing techniques to bi- and multi-static scenarios has not yet been fully realized with respect to the AUV. One of the greatest challenges for monostatic SAS imaging has been platform motion compensation. A theoretical framework for bistatic motion compensation has been recently developed using the so-called "memory line" of rough surfaces (Edwards and Schmidt, 2001), but has yet to be tested in a realistic environment. A greatly simplified, first order approximation of bistatic motion compensation with a stationary source was used in the GOATS'98 dataset, due to the lack of a synchronized source trigger. The GOATS 2000 dataset provides an opportunity to extend to full bistatic motion compensation with a stationary source. No dataset has yet been acquired for applying bistatic motion compensation with a moving source.

Another important application of bistatic scenarios is in the measurement of aspect-dependent target signatures. A stationary horizontal line array in the GOATS'98 experiment has been used to demonstrate the possibility of enhanced detection and classification with an appropriate bistatic configuration. Figure 5 clearly shows the preferential radiation of a flush-buried, fluid-filled cylinder in the GOATS'98 experiment. This regular radiation pattern provides clues to be applied for classification.

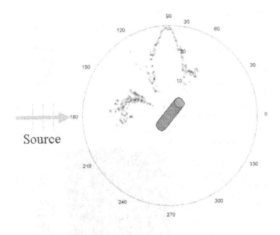

Source

Figure 5. Aspect-dependent radiated field of a flush-buried cylinder, as measured by a stationary horizontal line array. The signal measured at 90° from the source direction is 15 *dB* higher than the backscattered signal. The shape of the radiation pattern also provides classification clues.

Due to limited resolution capability at the low frequencies required for seabed penetration, alternative information processing techniques are also being explored to maintain the significant classification clues while coherently combining as much information as possible. One such application has utilized wavelet packets for detecting and individually filtering direct and elastic returns (Edwards and Montanari, 2001). This allows the different types of returns to be coherently combined in appropriate frequency bands and for the elastic returns to be associated with their respective direct returns.

4.   CONCLUSION

The GOATS Joint Research Program provides a series of coordinated, incremental implementations of the Autonomous Ocean Sampling Network concept for coastal REA and MCM. As the first of a series of experiments exploring new sensor concepts for the AOSN, GOATS'98 provided a unique dataset for developing new low-frequency, bistatic synthetic aperture processing approaces for mine countermeasures in very shallow water. As demonstrated here, such approaches has significant potential for detection of buried object beyond the critical bottom penetration range of traditional high-frequency MCM sonar systems.

Acknowledgments

The authors are highly appreciative of the collaboration with a large group of scientists and engineers from SACLANTCEN and MIT participating in the GOATS'98 experiment, and the crews of R/V Alliance and T/B Manning. This work is supported by the Office of Naval Research and SACLANT.

References

Curtin, T., Bellingham, J.G., Catipovic, J., and Webb, D. (1993). Autonomous oceanographic sampling networks. *Oceanography*, 6(3):86–94.

Edwards, J. R., Schmidt, H., and LePage, K. (2001). Bistatic synthetic aperture target detection and imaging with an AUV. *IEEE Journal of Oceanic Engineering*, 26(4):690–699.

Edwards, J. R. and Montanari, M. (2001). A wavelet packet approach for bistatic buried target classification with an AUV-borne synthetic aperture sonar. In *OCEANS 2001: An Ocean Odyssey*, Honolulu, HI.

Edwards, J. R. and Schmidt, H. (2001). Platform motion compensation for bistatic synthetic aperture sonar. In *Proceedings of the 17th International Congress on Acoustics*, International Commission for Acoustics, Rome, Italy.

LePage, K. D. and Schmidt, H. (2001). Bistatic synthetic aperture imaging of proud and buried targets from an auv. *IEEE Journal of Oceanic Engineering*, Submitted.

Maguer, A., Fox, W. L., Schmidt, H., Pouliquen, E., and Bovio, E. (2000). Mechanisms for subcritical penetration into a sandy bottom: Experimental and modeling results. *J. Acoust. Soc. Am.*, 107(3):1215–1225.

Schmidt, H., Edwards, J. R., LePage, K. D., and Bovio, E. (2000). Bistatic synthetic aperture sonar concepts for mcm AUV networks. In *Proceedings, International Workshop on sensors and sensing technologies*. Autonomous Undersea Systems Institute, Kona, Hawaii.

Schmidt, H. and Bovio, E. (2000). GOATS: Autonomous vehicle networks. In *Proceedings, Intl. Conf. Maneuvering and Control of Marine Craft*. International Federation for Automation and Control, Aalborg, Denmark.

Schmidt, H. and Lee, J. (1999). Physics of 3-d scattering from rippled seabeds and buried targets in shallow water. *J. Acoust. Soc. Am.*, 105:1605–1617.

DESIGN OF THE UMN MULTI-ROBOT SYSTEM *

Andrew Drenner, Ian Burt, Brian Chapeau, Tom Dahlin, Bradley
Kratochvil, Colin McMillen, Brad Nelson, Nikolaos Papanikolopoulos,
Paul E. Rybski, Kristen Stubbs, David Waletzko, Kemal Berk Yesin
Center for Distributed Robotics,
University of Minnesota, Minneapolis, MN 55455
corresponding author: Nikolaos Papanikolopoulos, npapas@cs.umn.edu

Abstract   Robotic reconnaissance and search and rescue are daunting tasks, especially in unknown and dynamic environments. The Scout is a robotic platform that is robust and flexible to operate in adverse and changing situations without revealing itself or disturbing the environment. The Scout can complete these missions by utilizing its small form factor for effective deployment, placement, and concealment while being equipped with a variety of sensors to accommodate different objectives. Unfortunately, the Scout has a limited volume to share among power, locomotion, sensors, and communications. Several novel approaches addressing deficiencies in specific tasks have been implemented in specialized Scouts and will be discussed in this paper. By building a diverse team of specialized Scouts, the team's strengths outweigh an individual weakness.

1.   INTRODUCTION

A robot platform that is designed for covert surveillance and reconnaissance will encounter a variety of obstacles in which a single robot or single type of robot may not be able to accomplish the objective of the mission within the parameters of the mission. These obstacles can take the form of limited passageways, non-traversable surfaces, limitations in time to complete a task, or operating in adverse or dynamic environments. Larger robots, which may be able to adapt to other situations and carry large payloads, are unable to easily conceal themselves in many environments nor is it possible for them to traverse small passageways. One method of dealing with these is to develop a team of

---

*This material is based upon work supported by the Defense Advanced Research Projects Agency, Microsystems Technology Office (Distributed Robotics), ARPA Order No. G155, Program Code No. 8H20, issued by DARPA/CMD under Contract #MDA972-98-C-0008.

A.C. Schultz and L.E. Parker (eds.), Multi-Robot Systems: From Swarms to Intelligent Automata, 141-148.
© 2002 *Kluwer Academic Publishers. Printed in the Netherlands.*

specialized small scale robots that can distribute a task rather than relying on a single larger robot.

Unfortunately, as the robots are made smaller, there is less room on-board to accommodate different environments or additional sensors. The Scout robot, developed at the University of Minnesota's Center for Distributed Robotics has been designed so that a team of specialized Scouts can work together to accomplish the tasks of semi-autonomous surveillance, reconnaissance, or search and rescue. This paper presents a reconnaissance scenario that utilizes the strengths of the members of the team to accomplish what no single member of the team would be able to. The specializations of the team involve changing both the locomotion methods of the Scout as well as the addition or replacement of hardware to improve the sensor capabilities of the Scout. This is followed by a look at some related work and finishes with a look at what can be done to further improve the Scout.

2.   A SAMPLE RECONNAISSANCE MISSION

The Scout robot was initially designed to be launched into a building to perform reconnaissance so that approaching individuals will know whether the building was occupied or could be considered safe. This objective expanded into a reconnaissance task which consisted of searching a collapsed or damaged building to find survivors rather than risk the lives of human rescuers or rescue dogs.

The traditional Scout robot, currently in its second generation is a small two-wheeled cylindrical robot, 40 mm in diameter and 110 mm in length. Shown in Figure 1, the Scout is capable of moving on relatively even surfaces on its wheels at approximately 0.31 m/s or hopping over obstacles through the use of its spring foot. The general Scout platform contains a variety of sensors including a black and white video camera, accelerometers, tiltometers, and wheel encoders. For more information on the general Scout platform, one may see (Hougen et al., 2000).

The general Scout fares well in search and rescue scenarios in which there are few obstacles and relatively even terrain such as an office building that may not have much structural damage. However, the low ground clearance of the Scout, approximately 3.2 mm, results in difficulty when crossing common office objects which may be on the floor such as cords or even pens and pencils. In situations in which the Scout is inhibited from forward rolling movement, the use of the spring foot enables it to hop in an arc over obstacles approximately 22 cm in height.

Figure 1. The Scout robot shown next to a CD for scale.

2.1 Debris Covered Surfaces

The typical Scout traverses flat surfaces and can overcome some small amounts of debris through its hopping mechanisms. However, a surface that is covered with debris can stop a traditional Scout in its tracks. The Actuating Wheel Scout has the capability of re-sizing the diameter of its wheels dynamically in the field which allow it to increase the ground clearance from the approximate 3.2 mm of the normal Scout to approximately 41.3 mm. This additional ground clearance allows the Scout to easily traverse much larger obstacles as seen in Figure 2. The improved ground clearance reduces the average speed when the wheels are at their largest diameter on smooth terrain to 0.2 m/s from the general Scout's 0.31 m/s to keep the torque of the smaller wheels.

There were several design criteria to be met by the actuated wheel system. The cylindrical form factor had to be maintained and the small size of the Scout required that no additional motors be added to directly power the actuation. The wheels needed to expand to at least twice the retracted size allowing for improved ground clearance. The wheels had to dynamically adjust to their environment, which is not possible with simple spring force. Finally, the wheel system needed to be as lightweight as possible yet still retain the strength needed to operate in adverse environments.

Of the possible designs, the current design has proven to closely match the specifications. The design involves the novel application of a latching solenoid to selectively couple the center wheel shaft to the body of the Scout. The inner side of the wheel is directly attached to the drive gear as well as having a bearing over the center shaft. The other side of the wheel is on a thread that runs along the center shaft. Thus, the drive motor is used to power the linear actuator in the wheel whenever the solenoid is engaged.

Figure 2.   The Actuating Wheel Scout crawling over debris.

Figure 3.   The Grappling Hook Scout carries a spring-loaded grappling hook that it can use to climb large obstacles.

2.2   Large Obstacle Navigation

When searching for survivors, rolling and hopping provide excellent forms of locomotion, but there are times when being so low to the ground limits what the Scout can see, especially when dealing with large obstacles such as desks, toppled file cabinets, or large debris. Rather than taking the time to drive around these large obstacles, the possible use of them as a vantage point for further investigation makes them extremely attractive. Unfortunately, the task of surmounting large obstacles is not something that a general Scout or one with actuating wheels is capable of, thus the need for the grappling hook Scout.

The grappling hook Scout, shown in Figure 3 is designed to provide an alternative method for crossing difficult terrain. With the grappling hook, a Scout can elevate itself onto a table, chair, bookshelf, or by using any other large object as an anchoring point. From its new vantage point, the Scout gains an improved range of vision and the possibility for a more concealed vantage point. Improved height also helps to reduce the effects of ground signal propagation and results in longer transmission distances. For further background on the grappling hook, one can see (Drenner et al., 2002).

2.3   Aerial Reconnaissance

Through the use of the actuating wheels and grappling hook, the Scout has a wide range of terrain that it can successfully conduct reconnaissance in. However, there are still areas which can not be overcome through rolling or climbing. In these situations, the application of a Scout controlled blimp offers the advantages of being able to control flight through an area with many large obstacles, an improved "bird's eye view" vantage point, and possible platform payloads such as additional lighting for darkened areas.

The blimp, shown in Figure 4 is controlled through an interface to the Scout. The Scout is capable of turning on and off as well as reversing the direction of a total of three fans. Two forward facing fans are used for going forward, reversing, and changing direction while a third fan is used to control lift. The blimp can be controlled to navigate corridors or stairwells, or hover in a stationary position for observation.

There are tradeoffs for using the blimp. The volume of the blimp is dependent upon the weight of the Scout and any other additional payloads that it may carry. The weight of a Scout and basic fans for control calls for a blimp with a volume of 0.4 m^3 of helium. This in turn requires that there is a minimum amount of clearance for the blimp Scout to pass through an area which determines whether it can make it through doors and around corners.

Figure 4. A Scout-controlled blimp. The fans that control the blimp's motion are actuated instead of the Scout's own wheels.

Figure 5. The IR Scout is equpped with infra-red emitting LEDs which allow the Scout's camera to be used in complete darkness.

2.4 Low Light Operation

In terms of reconnaissance for survivors, a key aspect for completing such a mission is to be able to function in situations where lighting may be less than ideal. In damaged buildings there may be rooms where there are no lights, or there may be situations where the site is too remote for immediate repair of electrical lines.

While originally intended as a method of supplying light while concealing the presence of the Scout in surveillance or other types of reconnaissance roles, an attempt to address low light operation requirements has been to add a pair of infrared (IR) emitters, each consisting of 36 IR diodes and a supplemental external battery pack to a Scout as shown in Figure 5. The extra battery pack requires that stronger, larger wheels replace the standard foam ones in addition to the gearing down of the drive motors. The standard black and white camera on the Scout is capable of using this additional illumination to navigate in low light areas.

A series of tests conducted using the autonomous behaviors in the software control architecture (Rybski et al., 2001) have shown that the camera currently in the Scout can be used to identify features with a 2 m radius. An additional benefit is that these IR emitters make it possible for other robots to observe the area of interest as well, thus only a small segment of the team will require this enhancement to facilitate low or no light operation.

2.5   Human Identification

Once a robot is in an area where there may be survivors, there are numerous ways to detect whether someone is there or not. Audio cues from survivors shouting from help or visual signs of motion can indicate that a survivor is near. However, when a survivor is knocked unconscious, these signs are not as readily available.

To improve the chances of finding survivors in these situations, certain Scouts have had been equipped with a color camera which replaces their standard black and white camera. The color cameras allow for video to be sent back and analyzed for skin tones which may identify potential survivors.

While the color camera is not that sensitive to the additional illumination offered from the IR emitters of the IR Scout, the newer cameras can operate with a lower amount of illumination than the unaided black and white ones. The new color cameras also offer improved resolution. The new cameras do have one drawback though, in that they consume more power than their predecessors.

3.   RELATED WORK

Small scale robots face an imposing world which require them to traverse hostile environments, adapt to dynamic situations, and interact with their environments to complete a variety of tasks. Combining the capabilities to complete these tasks within the constraints of the small form factor results in some rather remarkable robotic platforms.

Small robots often face a more difficult task when traversing hostile environments. Numerous forms of locomotion have been developed for small platforms, among them using a single gyroscopically stabilized wheel (Brown and Xu, 1997) for rolling, hopping systems for mobility in rough terrain (Hale et al., 2000), and self-reconfiguring robots such as Polybot (Yim et al., 2000) which is capable of reconfiguring itself to move like a snake, walk as a hexapod, or roll like a wheel.

Small scale autonomous flight is an interesting form of locomotion that has a very useful future. There are numerous larger fixed-wing unmanned aerial vehicles, but few have the ability to navigate indoors and none seem to be as adaptable as the Entomopter (Michelson and Reece, 1998). The Entomopter's insect like flight is suitable for flight not only in military reconnaissance roles, but also allows operation in thinner atmospheres making it ideal for Martian exploration.

Sensing the environment is very important for robots of any scale. In terms of urban search and rescue (Murphy et al., 2000), some of the most important sensors are those related to identifying victims such as thermal cameras and sensitive microphones. Sensors can also monitor for gas leaks or aftershock vibrations which can aid in the safety of rescuers.

Several projects have been focused on packing as many sensors into the smallest package possible to complete a mission. The Millibots (Bererton et al., 2000) are an example where small scale robots are combined with reconfigurable sensor packages. The mapping and surveillance capabilities of the Millibots are enhanced through the use of redundant sensors which address the shortcomings of a single sensor.

4. CONCLUSIONS AND FUTURE WORK

The Scout provides a unique and adaptable robotic platform capable of utilizing the diversity of the members of its team to accomplish difficult tasks in non-ideal conditions. Each Scout is suited for different operating conditions and teams of Scouts can be selected to handle specific situations depending upon the parameters of the current mission. In the example of searching a building for survivors, the Scout is able to traverse a variety of terrain, yet be small enough to drive through small holes in the walls without further damaging the surroundings.

Future generations of the Scout will have additional improvements to the mobility, communications, and sensing capabilities of the Scouts depicted here. Improvements in mobility will geared toward reducing the size further while still allowing for the freedom of movement found in devices such as the actuating wheels and grappling hook. Other sensors will be developed and integrated allowing for a wider variety of information retrieval. Increased communication range will allow wider team dispersal and the ability to operate in less wireless friendly environments.

The improvements come always in the form of a tradeoff, whether increasing mobility means a change in size or increasing sensing reduces overall operating lifetime by consuming more power. However, the incorporation of several specialized robots offsets the effects of the tradeoff because diversity in the team provides the strength to be both flexible and adaptable allowing for a wider range of missions and objectives than a single robot could accomplish.

References

Bererton, C., Navarro-Serment, L., Grabowski, R., Paredis, C. J., and Khosla, P. K. (2000). Millibots: Small distributed robots for surveillance and mapping. In *Government Microcircuit Applications Conf.*, Anaheim, CA.

Brown, H. and Xu, Y. (1997). A single-wheel gyroscopically stabilized robot. *IEEE Robotics and Automation Magazine*, 3(4):39–44.

Drenner, A., Burt, I., Dahlin, T., Kratochvil, B., McMillen, C., Nelson, B., Papanikolopoulos, N., Rybski, P. E., Stubbs, K., Waletzko, D., and Yesin, K. B. (2002). Mobility enhancements to the scout robot platform. In *Proc. of the IEEE Int'l Conf. on Robotics and Automation*, Washington DC, USA.

Hale, E., Schara, N., Burdick, J., and Fiorini, P. (2000). A minimally actuated hopping rover for exploration of celestial bodies. In *Proc. of the IEEE Int'l Conf. on Robotics and Automation*.

Hougen, D. F., Bonney, J. C., Budenske, J. R., Dvorak, M., Gini, M., Krantz, D. G., Malver, F., Nelson, B., Papanikolopoulos, N., Rybski, P. E., Stoeter, S. A., Voyles, R., and Yesin, K. B. (2000). Reconfigurable robots for distributed robotics. In *Government Microcircuit Applications Conf.*, pages 72–75, Anaheim, CA.

Michelson, R. C. and Reece, S. (1998). Update on flapping wing micro air vehicle research: Ongoing work to develop a flapping wing, crawling entomopter. In *13th Bristol International RPV Conference*, Bristol, England.

Murphy, R., Casper, J., Hyams, J., Micire, M., and Minten, B. (2000). Mobility and sensing demands in usar (invited). In *IECON*, Nagoya, Japan.

Rybski, P. E., Stoeter, S. A., Gini, M., Hougen, D. F., and Papanikolopoulos, N. (2001). Effects of limited bandwidth communications channels on the control of multiple robots. In *Proc. of the IEEE/RSJ Int'l Conf. on Intelligent Robots and Systems*, pages 369–374, Hawaii, USA.

Yim, M., Duff, D. G., and Roufas, K. D. (2000). Polybot: a modular reconfigurable robot. In *Proc. of the IEEE Int'l Conf. on Robotics and Automation*.

SIMULATING SELF-ORGANIZATION WITH THE DIGITAL HORMONE MODEL

Wei-Min Shen and Cheng-Ming Chuong
University of Southern California
4676 Admiralty Way, Marina del Rey, CA 90292
shen@isi.edu,chuong@pathfinder.usc.edu

Abstract   Inspired by the recent biological findings that many complex biological patterns are results of self-organization of homogeneous cells regulated by hormones, this paper presents the Digital Hormone Model (DHM) as a new computational model for self-organization. The model integrates advantages from many existing computational models of self-organization and produces results that match and predict the actual findings in the biological experiments of feather bud formation among uniform skin cells. To the best of our knowledge, these results are perhaps the first of its kind in the history of modeling self-organization. Furthermore, the new modeling techniques used in this model may lead more breakthroughs in simulating large scale and complex self-organization phenomena in nature.

Keywords: Multi-robot systems, self-organization, digital hormone model

1.   INTRODUCTION

The objective of this paper is to build upon the most recent biological experimental results and develop a general computational model for self-organization. In particular, we propose the Digital Hormone Model (DHM) that is generalized from an existing distributed control system for self-reconfigurable robots (Shen et al., 2000a; Shen et al., 2000b; Salemi et al., 2001). The model is inspired by the fact that many complex patterns in biological systems appear to be the results of self-organization among homogeneous cells regulated by hormones, and self-organization is based on local interactions among cells rather than super-imposed and pre-determined global structures (Jiang et al., 1999; Chuong et al., 2000). The DHM model integrates advantages from many existing modeling techniques for self-organization. This chapter describes the model in detail, reports the experimental results in simulating feather buds formation among homogeneous skin cells, and finds a number of correlations between individual cell/hormone profiles and the features of final patterns. These

A.C. Schultz and L.E. Parker (eds.), Multi-Robot Systems: From Swarms to Intelligent Automata, 149-157.
© *2002 Kluwer Academic Publishers. Printed in the Netherlands.*

results match the findings in the actual biological experiments and predict cases that have yet been observed in biological experiments but consist with the expected behaviors of hormone-regulated self-organization.

2.   SELF-ORGANIZATION IN BIOLOGY

Self-organizational phenomena are ubiquitous in nature. They appear in physics, chemistry, and materials sciences (Walgraef, 1996). But perhaps the richest source for self-organizational phenomena is biological systems. One very interesting self-organization phenomenon in biological systems is the formation of feathers. In chicken, for example, feathers are developed from skin cells during an early development stage before they hatch from the eggs. Abstractly speaking, homogeneous skin cells first aggregate and form feather buds that have approximately the same size and space distribution. The feather buds then grow into different types of feathers depending on the region of the skin. Many earlier theories believed that the periodic patterns of the feather buds are formed by sequential propagation and orchestrated by some "key" skin cells. These key cells occupy at some strategically critical positions on the skin. They first command their neighbor cells to form one sequence of feather buds and then this sequence will propagate to form other sequences of periodic patterns. However, recent findings in biological experiments have challenged these theories. Using disassociated mesenchyme (i.e., skin cells before becoming feather buds) and an intact epithelium (a thin layer on which skin cells can move and aggregate), Chuong and his colleagues (Jiang et al., 1999) have constructed a reconstitution system in which all mesenchymal cells are scrambled and reset to an equivalent state and have the same probability to become primodia or interprimodia (the feather buds). Surprisingly, the cells in this reconstitution system still grow into patterns of feather buds, and such growths occur almost simultaneously. These findings uncoupled the periodic pattern formation from the sequential propagation, and they suggest that there are no predetermined molecular addresses, and the periodic patterning process of feather morphogenesis is likely a self-organizing process based on physical-chemical properties and reactions between homogeneous cells.

Figure 1 shows the process of feather formation experiments in detail (Jiang et al., 1999; Chuong et al., 2000). During these experiments, the biologists also observed a set of interesting relations among the density of cell population, the diffusion characteristics of the activator and inhibitor hormones, and the size and the space distribution of the final feather buds. In particular, they observed that while the number of formed feather buds is proportional to the cell population density, the size of the feather buds remains approximately the same regardless of different population densities. The size of the feather buds, however, is related to the diffusion profiles of the activator and inhibitor

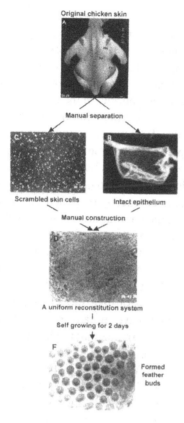

Figure 1. Self-organization in feather formation.

hormones. If the ratio of activator to inhibitor is high (low), then the final size of feather buds will be larger (smaller) than usual.

3. COMPUTATIONAL MODELS OF SELF-ORGANIZATION

Through out the history of science, there have been many computational models for self-organization. One of the earliest is perhaps Turing's morphogenesis model (Turing, 1952) in 1952, in which he analyzed the interplay between the diffusions of reacting species and concluded that their nonlinear interactions could lead to the formation of spatial patterns in their concentrations. Turing's model uses a set of differential equations to model the periodic pattern formation in a ring of discrete cells or continuous tissues that interact with each other through a set of chemicals he called "morphogens." Assuming that there are $r = (1, N)$ cells in the ring, and two morphogens X and Y among these cells. Let the concentration of X and Y in cell r be X_r and Y_r, the cell-to-cell diffusion rate of X and Y be u and v, and the increasing rate of X and Y caused by chemical reactions be $f(X, Y)$ and $g(X, Y)$, respectively, Turing modeled the dynamics of this ring as a set of differential equations and illustrated that a given ring of cells that initially have the uniform concentration of Y and X can self-organize, through random fluctuations and chemical reactions, into a ring of periodic patterns in the concentration of Y.

Cellular Automata (CA) is another important modeling technique for self-organization (Gutowitz, 1991). The most famous illustration of self-organization using CA is perhaps the game of Life, where randomly distributed cells on a space of grids will live or die based on a set of very simple and deterministic rules and gradually grow into many different stable patterns. When rules of a CA have stochastic characteristics, then they could be ideal for simulating interactions among many autonomous elements that perceive and react to local information in the environment.

Amorphous computing is another interesting technique that can potentially be useful for modeling self-organization. In an amorphous system, a large number of irregularly placed asynchronous, locally interacting computing elements are coordinated by diffusion-like messages and behave by rules and state markers. These systems have already been applied to building engineered systems for elements to organize and behave in a priori intend (Abelson et al., 1999; Nagpal, 1999). Similar ideas may be directly applicable to self-organization (Wolpert, 1969), provided that they do not solely rely on the positional information of the self-organizing elements.

Pheromone-based multi-agent systems are also of interesting as a tool for studying self-organization. (Parunak, et al., 2001) has already done a number of experiments to show that a set of autonomous agents can use pheromones to form interesting and complex global behaviors. Such an idea shares many common design principles as digital hormones described here.

4.   THE DIGITAL HORMONE MODEL

Inspired by the biological findings described above, we propose the Digital Hormone Model (DHM) to integrate existing computational models for self-organization, and generalize an earlier distributed control system for self-reconfigurable robots (Shen et al., 2000a; Shen et al., 2000b; Salemi et al., 2001).

A digital hormone model for self-organization consists of a space of grids and cells. Among these grids, cells can live, evolve, migrate, or die as time passes. Each living cell occupies one grid at a time and a cell can secrete chemical signals (hormones) to diffuse into its neighboring grids to influence other cells' behaviors. Two types of hormones are most common: an activator hormone will encourage certain reactions in cells, while an inhibitor hormone

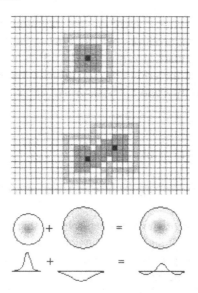

Figure 2.   A simple DHM with two actions and two hormones.

will prohibit certain cell reactions. Every hormone diffuses to its neighboring grids according to its own diffusion function.

Each cell is an autonomous and "intelligent" entity and can react to hormones and perform a set of possible actions: migration to a neighbor grid, secretion hormones, change cell features, proliferation of cell, death of cell, and adhesion to connect neighbor cells. At any given time, a cell selects and executes one or more actions according to a set of internal behavior rules. In this paper, we assume that the rules are given, unchangeable, and consistent. At any time, the rules do not cause a cell to select actions that are conflict to each other. Given the grids, cells, hormones, actions, and rules, the Digital Hormone Model works as follows:

1 All cells select actions by their behavior rules;

2 All cells execute their selected actions;

3 All grids update the concentrations of hormones based on their diffusion functions.

4 Go to 1.

To illustrate the above definitions, let us consider a simple DHM in Figure 2, where cells (marked as the black dots) migrate on a space of NxN grids. The space is circular in the sense that the leftmost and the rightmost columns are neighbors, and the topmost and the bottommost rows are neighbors. Cells in this DHM have only two actions: secretion and migration, and the former is a constant action that always produces two hormones: the activator A (marked as red) and the inhibitor I (marked as green). The diffusion rates for A and I are also constant. For a cell at the grid (a,b), the diffusion rates of A and I around the cell are characterized by s and r, respectively, where $s < r$, and they obey the standard distributions:

$$f_A(x,y) = (2\pi s^2)^{-1} e^{[(x-a)^2+(y-b)^2]/2s^2}$$

$$f_I(x,y) = -(2\pi r^2)^{-1} e^{[(x-a)^2+(y-b)^2]/2r^2}$$

Notice that because $s < r$, the hormone A has a sharper and narrower distribution then the hormone I, and these characteristics are similar to those observed in the biological experiments. We assume that the two hormones are combinable so that the concentration of hormones in any given grid can be computed by summing up all present "A"s and "I"s in the grid. Thus, for a single isolated cell, the concentration of hormones around its grid will be the sum of A and I and will form three "colored" rings (see the low part in Figure 4). The activator hormone dominates the inner ring (the red); the inhibitor hormone dominates the outer ring (the green); and the middle ring is neutral (the white grids). When two or more cells are near each other, the hormones in the surrounding grids are summed up to compute the hormone strengths. In the upper part of Figure 4, we have illustrated the combined hormones around a single cell and around two nearby cells. Since the grids are discrete, the rings around cells are shown as squares.

In this simplified DHM, two simple rules govern the actions. One rule states that "secret A and I for every step", and the other states that "migrate to a neighbor grid based on hormone distribution." More specifically, the probability for a cell to migrate to a particular neighboring grid (including the current occupying grid) is proportional to the concentration of A and inverse proportional to the concentration of I in that grid. This rule is fundamentally stochastic, so the selection of migrating grid is non-deterministic. The collisions of cells are solved in a synchronized manner. All cells first "virtually" move to the grids they selected. If there are multiple cells in the same grid, then the grid will randomly distribute the extra cells to those immediate neighboring grids that are empty. This ensures that there is no more than one cell in each grid at any time.

5.   EXPERIMENTAL RESULTS WITH DHM

In this paper, we would like to use the DHM to test the following hypotheses: (1) Will the digital hormone model enable cells to self-organize into any patterns at all? (2) Given the same hormone diffusion profiles, will the size of final patterns be invariant to the cell population and density, so as to match the observations made in the biological experiments? (3) How will the hormone diffusion profiles affect the self-organization process and the size of the final patterns? (4) Will arbitrary cell/hormone profile enable self-organization and allow cells to form patterns?

To test these hypotheses, we have run two sets of experiments using the simplified digital hormone model described in Figure 2. In the first set of experiments, we set the hormone diffusion profile as follows to approximate the standard distributions. For any single isolated cell, let the cell's nth ring of neighbors be the neighboring grids at the distance of n grids away from the cell. Using this definition, we define the concentration level of the activator hormone at the cell's surrounding grids as follows: 0.16 for the 0th ring (i.e., the occupying grid), 0.08 the 1st ring, 0.04 the 2nd ring, 0.02 the 3rd ring, and 0 the 4th and beyond. For the inhibitor hormone, the concentration levels for the 0th through the 4th rings of neighbors are: -0.05, -0.04, -0.03, -0.02, and -0.01. Thus the combined concentration levels of hormones at these rings are: 0.11, 0.04, 0.01, 0, and -0.01. We assume that the concentrations of hormones secreted by a cell at grids beyond the 4th ring are so insignificant that they can be practically ignored.

Given this fixed hormone diffusion profile, we have run a set of simulations on a space of 100x100 grids with different cell population densities ranging from 10%, 25%, through 50%. Starting with cells randomly distributed on the grids, each simulation runs up to 1,000 action steps, and records the configuration snapshots at the steps of 0, 50, 500, and 1,000. As we can see from the results in Figure 3, cells in all simulations indeed form clusters with approximately the same size. These results demonstrate that the digital hormone model indeed enables cells to form patterns. Furthermore, the results match the observations made in the biological experiments. The size of the final clusters does not change with the cell population and density, but the number of clusters does. Lower cell densities result in less numbers of final clusters, while higher densities form more clusters.

In the second set of experiments, we have varied the hormone diffusion profiles and observed their effect on the results of pattern formation. The results are shown in Figure 4. As we can see, when a balanced profile of activator and inhibitor is given (see the second row), the cells will form final patterns as before. As the ratio of activator over inhibitor increases (see the third row), the size of final clusters also increases. These results match exactly with the findings in biological experiments.

When the ratio of A/I becomes so high that there are only activators and no inhibitors (see the fourth row), then the cells will form larger and larger clusters, and eventually become a single connected cluster. On the other hand,

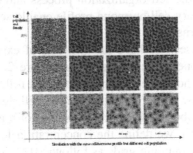

Figure 3.   Simulation with the same cell/hormone profile but different cell population.

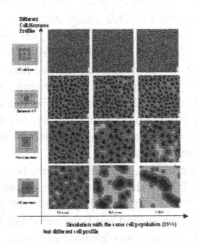

Figure 4.   Simulation with the same cell population (25%) but different cell profiles.

when the ratio is so low that there is only inhibitor and no activator, then the cells will never form any patterns (see the first row), regardless how long the simulation runs. These results are yet to be seen in biological experiments, but they are consistent with the principles of hormone-regulated self-organization, and well qualified as meaningful predictions of cell self-organization by hormones.

Acknowledgments

We are grateful that this research is in part supported by the AFOSR contract F49620-01-1-0020 and the DARPA contract DAAN02-98-C-4032.

References

Abelson, H., Allen, D., Coore, D., Hanson, C., Homsy, G., Knight, T. F., Nagpal, R., Rauch, E., Sussman, G. J., and Weiss, R. (1999). *Amorphous Computing,* The MIT Press, Cambridge, MA.

Chuong, C.-M., Chodankar, R., Widelitz, R. B., and Jiang, T. X. (2000). Evo-Devo of Feathers and Scales: Building complex epithelial appendages. *Current Opinion in Development and Genetics,* 10: 449-456.

Gutowitz, H., editor. (1991). *Cellular Automata — Theory and Experiment,* The MIT Press, Cambridge, MA.

Jiang, T. -X., Jung, H. S., Widelitz, R. B., and Chuong, C. -M. (1999). Self organization of periodic patterns by dissociated feather mesenchymal cells and the regulation of size, number and spacing of primordia. *Development* 126:4997-5009.

Nagpal, R. (1999). *Organizing a global coordinate system from local information on an amorphous computer,* MIT, Boston.

Parunak, H.V.D. and Brueckner, S. (2001). *Entropy and Self-Organization in Multi-Agent Systems.* In *International Conference on Autonomous Agents,* Montreal, Canada.

Salemi, B., W. -M. Shen, and Will, P. (2001). Hormone-Controlled Metamorphic Robots. In *Proceedings of the International Conference on Robotics and Automation,* Seoul, Korea.

Shen, W. M., Lu, Y., and Will, P. (2000a). Hormone-based Control for Self-Reconfigurable Robots. In *Proceedings of the International Conference on Autonomous Agents,* Barcelona, Spain.

Shen, W. M., Salemi, B., and Will, P. (2000b). Hormones for Self-Reconfigurable Robots. In *Proceedings of the 6th International Conference on Intelligent Autonomous Systems,* Venice, Italy.

Turing, A.M. (1952). The chemical basis of morphogenesis. *Philos. Trans. R. Soc. London B,* 237:37-72.

Walgraef, D. (1996). *Spatio-Temporal Pattern Formation.,* Springer.

Wolpert, L. (1969). Positional information and the spatial pattern of cellular differentiation. *Journal of Theoretical Biology,* 25:1-47.

VI

DESIGN AND LEARNING

ARCHITECTING A SIMULATION AND DEVELOPMENT ENVIRONMENT FOR MULTI-ROBOT TEAMS

Stephen Balakirsky, Elena Messina, James Albus
National Institute of Standards and Technology
Intelligent Systems Division
Gaithersburg, MD 20899-8230
stephen@cme.nist.gov, elena.messina@nist.gov, albus@cme.nist.gov

Abstract   Collaborative multi-robot teams have great potential to add capabilities and minimize risk within the military domain. The composition of these teams may range from multiple copies of the same model to heterogeneous ground, air, and water vehicles operating in concert. The novel and extremely complex nature of these autonomous systems requires a large up-front investment in design, modeling, simulation, and experimentation. Myriad design decisions must be made regarding the control architecture and general information flow for commands and status exchanged between robots and humans and among robot teams. In addition, decisions must be made on how to best assemble and deploy these teams. To assist in making these design decisions, we are developing an integrated environment that we hope will greatly facilitate the design, development, and understanding of how to configure and use multi-robot teams for military applications and to accelerate the robots' deployment.

Keywords: simulation, architectures, 4-D/RCS, autonomous vehicles

1.   INTRODUCTION

Military operations are being redesigned to incorporate robotic vehicles. The recent successful missions flown by Predator and Global Hawk in Afghanistan have increased the visibility of, and confidence in, unmanned vehicles. The United States Army is transforming its combat forces to be lighter, more agile, and network centric. The Future Combat Systems (FCS) Program, run out of the Defense Advanced Research Projects Agency (DARPA) and jointly conducted with the U. S. Army will distribute the sensing, weapon delivery, and command and control elements (Gourley, 2000). A single manned vehicle will control a series of robotic weapons and sensors. The Army Research Laboratory's Demo III Program culminated in November 2001 with

A.C. Schultz and L.E. Parker (eds.), Multi-Robot Systems: From Swarms to Intelligent Automata, 161-168.

a demonstration of robotic scout vehicles performing missions in concert with soldiers (Murphy et al., 2002). Key technologies – both hardware and software – are becoming available to be exploited by autonomous vehicles.

Despite the early successes of unmanned aerial vehicles and of Demo III, there is still a tremendous amount of research, design, development, and integration work to be done. There are several layers of difficulty that must be tackled. First of all, building and validating software that works in concert with complex hardware and that achieves robust autonomy is a major challenge. Integration of multiple vehicles that cooperatively execute a mission is another level of difficulty. Cooperation among heterogeneous vehicles (ground, air, water) is yet a third level of challenge. Besides the technological challenges, there exists the need for the military doctrine to evolve to be able to fully exploit these new assets. These challenges cannot wait until there are live experiments with the new autonomous platforms. Not only would it be unrealistically costly, but it would delay evaluation and design decisions until it is too late to revoke them. In order to meet the revolutionary needs of the new armed forces, simulation facilities that are integrated with software development capabilities will be crucial elements at every stage of this process.

2.   AN ARCHITECTURE FOR INTELLIGENT AUTONOMOUS VEHICLES

One of the key decisions to be made in building any kind of complex system is how to organize the hardware and software. The Demo III Program and some of the teams competing for the Future Combat Systems contract have selected the 4-D/RCS reference architecture for their autonomous vehicles (Albus, 1999). The 4-D/RCS architecture consists of a hierarchy of computational nodes, each of which contains the same elements. Based on control theory principles, 4-D/RCS partitions the problem of control into four basic elements that together comprise a computational node: behavior generation (BG), sensory processing (SP), world modeling (WM), and value judgement (VJ). Figure 1 shows the 4D/RCS control node and the connections between its constituent components. Figure 2 shows a sample 4-D/RCS hierarchy for military scout vehicles.

3.   REQUIREMENTS FOR A SIMULATION, MODELING, AND DEVELOPMENT ENVIRONMENT

An architecture is a first step towards guiding and facilitating the construction of complex multi-vehicle autonomous systems. Tools that help automate the software development are another important element. NIST has been working with industry, other government agencies, and academia to investigate tools

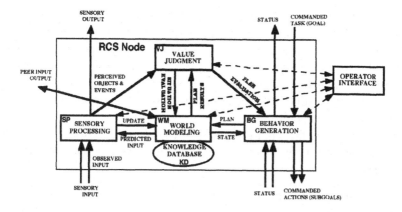

Figure 1.   Model for RCS Control Node.

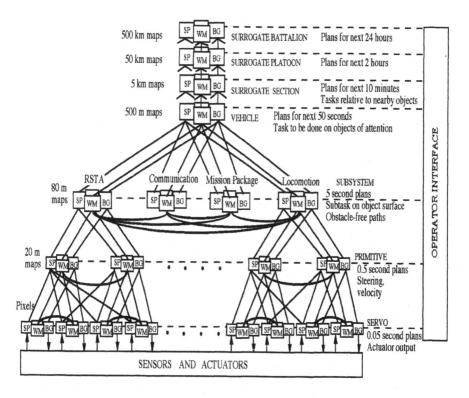

Figure 2.   4-D/RCS Reference Model Architecture for an individual vehicle.

to facilitate construction of the types of large and complex systems as needed by Demo III and FCS. We are developing a large-scale simulation environment that will enable us, along with others, to design the control hierarchy, populate the control nodes, run the system in simulation, debug it, and generate code for the target host. The development and simulation environments are closely tied to the eventual deployment platforms and are intended to be able to operate with a combination of real and simulated entities. The ability to enable human-in-the loop testing and execution is also crucial, given the novel aspects of human-robot interactions.

A high-level list of the requirements for such a development and simulation environment has been developed to help guide its creation. The requirements are as follows:

- Full support of 4-D/RCS architecture
- Graphical user interface for building, testing, debugging the system under development
- Reuse support:
    - Architecture elements
    - Component templates
    - Algorithms
    - Code
    - Subsystems
- Intuitive visualizations of the control system to support design and use and provide an understanding of what the system is doing and why, and what it plans to do next. Examples of visualizations include:
    - Display of control hierarchy as it executes, including commands flowing down and status flowing up
    - Ability to "zoom in" on a particular node and view states as the system executes
    - Ability to view world models within the system
- Execution controls, including
    - Single step through execution
    - Breakpoints and watch windows
    - Logging
- Simulation infrastructure supporting realistic execution scenarios, visualization, and debugging/evaluation experiments. This includes
    - Population of the environment external to the vehicle with relevant features (such as roads, other pieces of equipment, humans, etc.)
    - Dynamic environment (with moving entities)
- Modification capabilities so that the designer and user can perform "what if" experiments. The tools should allow interactive and intuitive modification of situations in the environment or within the system. The modi-

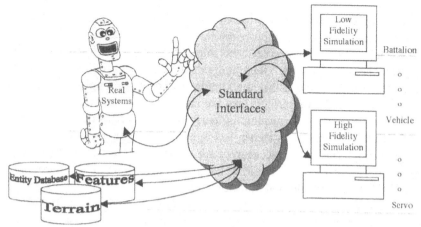

Figure 3. Hierarchy of simulators.

fication capabilities should work seamlessly with the visualization, simulation, and execution features. Examples of types of modifications that should be allowed include:

- Changing world model data
- Importing datasets that represent what the system's sensors would receive
- Changing environmental conditions

4. PROPOSED SYSTEM OF SYSTEMS

We are seeking to create an *integrated* environment that provides capabilities typically associated with software development tools and those associated with simulation environments. Whereas several commercial tools exist to help design and construct software, these tool-sets are disconnected from rich execution environments (real or simulated). There also exist many sophisticated simulation systems, but they work at either a very broad scale at low resolution, or at high resolution. What we are attempting is a totally coherent environment for designing, developing, and validating a team of vehicles. The software design development support aspects include being able to work in a graphical environment to sketch the control hierarchy, bring up partially-filled in 4-D/RCS control nodes, easily create connections between nodes (or components within them), and automatically generate executable code. The software support will also encompass capabilities typically found under run-time debug tools, including single stepping and setting break points.

The software development tools will segue smoothly into the simulation environment. Under this concept, a virtual world is being created that brings together existing multi-platform and single platform simulation systems into

a system of systems. Through the use of well defined interfaces that are supported on a wide variety of computer platforms, the simulator's internal command and data flows will be able to be interrupted and modified. This will
allow researchers to "plug-in" their individual technology components and
override the default methods that the simulators normally employ. Through
the use of these standard interfaces, we are attempting to provide researchers
with a low cost technique for evaluating performance changes due to their algorithm's implementation. As shown in Figure 3, interfaces will be provided
that range through the entire spectrum of the 4-D/RCS hierarchy; from a low-
fidelity multi-platform configuration to a high-fidelity single platform configuration, to the ability to add real platforms into the virtual world. In addition,
global variable resolution database resources will be offered. These include a
terrain database that contains elevation data, a feature database that contains
vector data for roads, buildings, rivers, etc., and an entity database that contains information on all of the platforms participating in the simulation. Filters
will be available to tune the database outputs to the specific needs of each algorithm. For example, a low-level mobility planner may require a 40 cm cell size
in an elevation array while a high-level planner may require a 40 m cell size.
In addition to serving a priori data, these databases will be able to be modified
in real-time. Any modifications made to the databases will be viewable by all
participants (both real and virtual) in the exercise.

At the top of the hierarchy, a low-fidelity, long temporal and spatial duration, multi-platform simulator will be used. As designed, this class of simulator
is capable of simulating the interaction, coordination, and movement of large
groups of platforms. While these simulators do simulate down to the level
of the individual platforms moving across the terrain, the terrain and mobility
models are typically low resolution. Therefore, this class of simulator is best
utilized in developing algorithms for group behaviors where precise platform
mobility and sensing modeling is beyond the scope of the experiment. A second class of simulator will be employed for situations where precise modeling
is required. These simulators will share interfaces with the low-fidelity simulator, and in fact may take commands from the low-fidelity simulators in order to
precisely model one or more of the platforms involved in a particular exercise.
The high-fidelity simulators will be able to read the shared databases and construct simulated sensor output (or pass real sensor output) that may be used by
external sensor processing algorithms. Complex, dynamically correct platform
motion models and high resolution terrain information will also be available at
this level.

Interfaces will be inserted into each simulator that will enable the export and
import of both world model and behavior generation information at each level
of the 4-D/RCS hierarchy. This will enable researchers to implement a particular group behavior or algorithm at a particular level of 4-D/RCS. For example,

a cooperative search algorithm could be implemented at the "section" level of 4-D/RCS. The algorithm would receive its command input from the platoon level of the low-fidelity simulator and construct a plan based on information read from the terrain, entity, and feature databases. The planned course of action (individual plans for several platforms) would then be passed back into the simulator for execution. In this particular case, the plans could be passed either back to the low-fidelity simulator or to the high-fidelity simulator. In addition, one or more of the platform's plans could be passed to real systems. As designed, the source and destination of these plans and the data utilized to construct them is transparent to the planning system and totally controlled by the user. This will facilitate an environment where a researcher can simulate as many or as few subsystems and levels as desired.

5.   CURRENT IMPLEMENTATION

While the entire software development and simulation system has not yet been implemented, progress has been made on developing prototypes and designs for the overall system. A simulation testbed system has been built with emphasis placed on developing interfaces into the shared databases and into a low-fidelity simulator. These interfaces will allow our individual real robots to interact in simulated group behaviors, and our group behavior planning systems to plan for mixed groups of simulated/real robots. For the initial system, we have chosen the US Army STRICOM's OneSAF Testbed Baseline [1] for both the low-fidelity simulation and shared database server. We have worked closely with the Army Research Laboratory and Science and Engineering Services Inc. to install the standard interfaces. All of the interfaces communicate over NIST's NML communication channels (Shackleford et al., 2000) which provide a multi-platform solution to inter-process communication.

Distributed, shared databases are implemented as part of the standard implementation of OneSAF. We have added interfaces into the simulation system that allow for simple outside access of this information. These interfaces include hooks into the terrain elevation database, feature database, and entity database. For the terrain elevation database, both all-knowing (what is the elevation in this area, to this resolution) and modeled (what is the terrain map as modeled by vehicle x with its sensors) are available. Feature vector data is currently available only on an all-knowing basis. For entity data, filtered information (all friendly, enemy, detected, etc.) reports are available as well as event detections. Events currently supported include line crossings and anticipated line crossings with more to be added shortly.

---

[1] The identification of certain commercial products does not imply recommendation or endorsement by NIST.

In addition to the database access interfaces, we are able to interrupt the standard OneSAF command flow to inject our own plans. This has been demonstrated by having OneSAF section level plans sent out over an NML channel to a stand-alone vehicle level planner. The results of the vehicle level planner can then be executed on real robotic hardware, sent to a high-fidelity simulator, or sent back into the OneSAF simulator for execution.

Work is continuing on developing further interfaces. These new interfaces will include the ability for a real robotic platform to influence OneSAF databases by continuously updating their own location as well as adding detected features and entities. In addition, further breaks in the OneSAF command flow will be implemented to allow for planning systems that compute group plans to be implemented and evaluated.

In terms of the software development support, we have been experimenting with various representation techniques and development tools. These range from commercial packages, such as Real-Time Innovation Incorporated's ControlShell to novel formal languages, such as Stanford's Rapide (Messina et al., 1999). Recent work has focused on the use of the Unified Modeling Language to support 4-D/RCS control system development (Huang et al., 2001). A commercial development and execution tool for building simpler versions of RCS-style controllers has been developed by a small company (Advanced Technology and Research), but it is targeted at manufacturing systems that have minimal sensing requirements. We are currently working with outside partners to develop a prototype control system development tool that permits the types of visualizations, modifications, and execution controls as were listed above.

References

Albus, J. (1999). 4-D/RCS reference model architecture for unmanned ground vehicles. In Gerhart, G., Gunderson, R., and Shoemaker, C., editors, *Proceedings of the SPIE AeroSense Session on Unmanned Ground Vehicle Technology*, volume 3693, pages 11–20, Orlando, FL

Gourley, S. (2000) Future combat systems: A revolutionary approach to combat victory *Army*

Huang, H., Messina, E., Scott, H., Albus, J., Proctor, F., and Shackleford, W. (2001) Open system architecture for real-time control using an UML-based approach. In *Proceedings of the 1st ICSE Workshop on Describing Software Architecture with UML*

Messina, E., Dabrowski, C., Huang, H., and Horst, J. (1999) Representation of the rcs reference model architecture using an architectural description language. In *Lecture Notes in Computer Science EUROCAST 99*, volume 1798 of *Lecture Notes in Computer Science*. Springer Verlag

Murphy, K., Abrams, M., Balakirsky, S., Chang, T., Lacaze, A., and Legowik, S. (2002) Intelligent Control For Off-Road Driving. In *Proceedings of the First International NAISO Congress on Autonomous Intelligent Systems*

Shackleford, W. P., Proctor, F. M., and Michaloski, J. L. (2000) The neutral message language: A model and method for message passing in heterogeneous environments. In *Proceedings of the 2000 World Automation Conference*

ROBOT SOCCER:
A MULTI-ROBOT CHALLENGE

Manuela M. Veloso
School of Computer Science
Carnegie Mellon University
*Pittsburgh, PA 15213, USA**
veloso@cs.cmu.edu

Abstract   Robot soccer opened a new horizon for multi-robot research: Teams of autonomous robots need to respond to a highly dynamic and uncertain environment including other teams of robots. Furthermore, soccer robots have clear and
specific goals to accomplish. The multi-robot system relies both on robust autonomous individual robots and teamwork. We have developed new algorithms
for localization, navigation, and teamwork, to demonstrate real-time performance in this multi-robot adversarial task. Robot soccer has also a strong entertainment component attracting researchers and crowds of spectators. RoboCup-
2001, the international robotic soccer competitions held for the first time in the
United States in 2001, joined more than 500 participants, 200 robots, and a few
thousand spectators.

Keywords: Multi-robot adversarial environments, real-time autonomous robots

1.   INTRODUCTION

The late Herbert A. Simon, in the conclusion of his lecture at the Earth-
ware Symposium at Carnegie Mellon on "Forecasting the Future or Shaping
it?" (Simon, 2000) said: "Here around CMU, we have been amazed, amused,
gratified, and instructed by the developments in robot soccer. For four years,
and with rapidly increasing skill, computers have been playing a human game
requiring skillful coordination of all the senses and motor capabilities of each
player, as well as communication and coordination between players on each
team, and strategic responses to the moves of the opposing team. We have
seen in the soccer games, an entire social drama, played out with far less skill
(thus far) than professional human soccer, but with all the important compo-

*Partial funding provided by DARPA grants F30602-00-2-0549, DABT63-99-1-0013, and F30602-98-2-
0135. The content of this publication is the sole responsibility of the author.

A.C. Schultz and L.E. Parker (eds.), Multi-Robot Systems: From Swarms to Intelligent Automata, 169-174.
© 2002 *Kluwer Academic Publishers. Printed in the Netherlands.*

nents of the latter clearly visible. Here we see, in a single example, a complex web of all the elements of intelligence and learning – interaction with the environment and social interaction, use of language – that artificial intelligence has been exploring for half a century, and a harbinger of its promise for continuing rapid development. Almost all of our hopes and concerns for the future can be examined in miniature in this setting, including our own role in relation to computers."

Herb Simon's lecture goes on very interestingly forecasting our interactions with computers and robots. But his impressions and assessment of robotic soccer set a tremendous and exciting responsibility for our multi-robot research.

In robotic soccer, robots face a very dynamic and uncertain environment where they have to achieve very clear goals, such as score a ball into an opponent goal. Robotic soccer teams need to effectively integrate perception, action, and cognition in real-time. Each team of robots needs "to close the loop," as I always say. Robots need to continuously live in a cycle, perceiving the world, deciding what to do, and performing its actions. One of the main challenges has shown to be exactly this ability to close this "autonomy loop," namely to perceive the environment, make decisions about the actions to take, actually take actions in the world, and continuously again perceive the environment, make decisions, and act.

We have been developing several different teams of autonomous robot soccer players that offer different technical challenges of perception, action, and cognition.

2.   TEAMS OF ROBOTS

One of the main challenges of robotics is the integration of many research accomplishments into a single *complete* robot. Research remarkably advances in several separate directions, but a complete robot requires the integration of many different capabilities. In addition, there is a real research challenge on how to create *groups* of robots rather than single individual ones.

Actually developing real robots as opposed to simulation is far from trivial. One of the main bottlenecks is the robot's perception. Robots need to be able to accurately and reliably infer the state of the world.

Global Perception and Distributed Action.   The small-size robot soccer RoboCup league allows for global perception by a camera that view the complete field. Figures 1 and 2 show some of different teams of small-wheeled soccer-playing robots that we have developed at Carnegie Mellon (Veloso, et al., 1998; Veloso, et al., 1997).

Each participating team designs and builds their own robots under specific size constraints. These robots play on a field of approximately the size of a ping-pong table. They play with an orange golf ball. The robot teams are al-

Figure 1. Small-Wheeled Soccer Robots - Carnegie Mellon teams since 1997. Thanks to Sorin Achim, Michael Bowling, Kwun Han, and Peter Stone.

lowed to hang a vision camera over the playing field that has a global vision of the complete world. This global perception problem proved to be more complicated than it may seem. Indeed, processing images globally in real time with eleven moving objects at high speeds is in itself a real perception challenge. But, the processed images of the global view of the world are available to each robot in a team. The image can be passed to an off-board computer that remotely controls the robots usually through radio. Interestingly, because the robots now have a complete view of the positions of all the other teammates and opponents, they can effectively use this information to strategically collaborate as team members.

Figure 2. Small Soccer Robots - Carnegie Mellon team for 2001 and 2002. Thanks to Michael Bowling, Brett Browning, James Bruce, and Ravi Balasubramanian.

The robots achieve teamwork by playing different roles in the team. With five robots, we have developed behaviors for attacker, mid-fielder, defender, and goalie robots. The robots use reactive behaviors, mapping the state of the world at each moment and deciding which action to take. We have developed an algorithm that coordinates multiple attacking robots in which one robot goes to the ball and the other robots move to position themselves in an adequate open area to anticipate a possible pass. The robots run an objective function optimization, SPAR, that *strategically positions* robots to be *attracted* to the goal and ball (i.e., minimizing their distance to the goal and to the current position of the ball), and to be *repulsed* from the other robots (i.e., maximizing their distance to the other robots). The combination of role behaviors and SPAR allowed the robots to demonstrate effective teamwork.

On-board Perception, Cognition, and Action.   The Sony RoboCup legged league offers where robots are fully autonomous. The Sony robots have vision and computational power on board of the robots. Figures 3 and 4 shows the legged robots that we have used in the last years. We have used the Sony legged robots since their very first version in 1998 (Veloso, et al., 1998; Veloso, et al., 2000).

Figure 3.   Sony Legged Soccer Robots - The Carnegie Mellon 1999 and 2000 robots. Thanks to James Bruce, Scott Lenser, and Elly Winner.

The robots are autonomous, and have on-board cameras. The on-board processor provides image processing, localization and control. The robots are not remotely controlled in any way, and as of now, no communication is possible in this multi-robot system. The only state information available for decision making comes from the robot's on-board colored vision camera and from sensors which report on the state of the robot's body. The vision algorithm is hence of crucial importance as it provides the perception information as the observable state. Our vision system robustly computes the distance and angle of the robot to the objects and assigns confidence values to its state identifications (Bruce, et al., 2000).

Figure 4. Sony Legged Soccer Robots - The Carnegie Mellon 2001 robots. Thanks to James Bruce, Martin Hock, Scott Lenser, and Will Uther.

The preconditions of several behaviors require the knowledge of the position of the robot on the field. The localization algorithm is responsible for processing the visual information of the fixed colored landmarks of the field and outputting an (x, y) location of the robot. Because effective robots are small and given the dynamics of the game, the robots' motion can be affected by external sources beyond the robot's own control (e.g., referee, other robots' pushing, slanted walls). We developed a new sensor-resetting probabilistic localization algorithm was devised which allows robots to use their sensor input to rapidly adapt to changes in their position (Lenser and Veloso, 2000).

Finally, our behavior-based planning approach interestingly provides the robot the ability to control its knowledge of the world. Behaviors range from being based almost solely on the visual information to depending on accurate localization information. Furthermore behaviors vary as a function of the confidence of the robot in its world model (Winner and Veloso, 2000). The robots for RoboCup-2002 will be able to communicate between themselves. This will open a new opportunity for multi-robot coordination.

All the teams in the RoboCup legged robot league use this same Sony hardware platform. This creates a very interesting research question, as now all the robots have in principle the same perception and motion low-level capabilities. And therefore their eventual different performance should be mainly the cognition aspect. However, this is indeed not the case. Although they do differ at the cognition level, it is still a challenge to program the robots *to use* their similar hardware. So some robots move faster or see better than other robots.

3.   CONCLUSION

The examples of robot soccer illustrate the challenges of building complete autonomous robots that can perform active perception and sensor-based planning while playing a multi-robot game. The robot soccer games have shown to be not only a true source of entertainment but a great source of advances in robotics research.

References

Bruce, J., Balch, T., and Veloso, M. (2000). Fast and inexpensive color image segmentation for interactive robots. In *Proceedings of IROS-2000*, Japan.

Lenser, S., and Veloso, M. (2000). Sensor resetting localization for poorly modeled mobile robots. In *Proceedings of ICRA-2000, the International Conference on Robotics and Automation*.

Simon, H. A. (2000). Forecasting the Future or Shaping it? *Earth-ware Symposium at Carnegie Mellon*, Lecture, see video at http://www.ul.cs.cmu.edu/.

Veloso, M., Bowling, M., Achim, S., Han, K., and Stone, P. (1999) The CMUnited-98 champion small robot team. In Minoru Asada and Hiroaki Kitano, editors, *RoboCup-98: Robot Soccer World Cup II*, pages 77–92. Springer.

Veloso, M., Stone, P., and Han, K. (1998). CMUnited-97: RoboCup-97 small-robot world champion team. *AI Magazine*, 19(3):61–69.

Veloso, M., Uther, W., Fujita, M., Asada, M., and Kitano, H. (1998). Playing soccer with legged robots. In *Proceedings of IROS-98, Intelligent Robots and Systems Conference*, Victoria, Canada.

Veloso, M., Winner, E., Lenser, S., Bruce, J., and Balch, T. (2000). Vision-servoed localization and behavior-based planning for an autonomous quadruped legged robot. In *Proceedings of the Fifth International Conference on Artificial Intelligence Planning Systems*, pages 387–394, Breckenridge, CO.

Winner, E., and Veloso, M. (2000). Multi-fidelity behaviors: Acting with variable state information. In *Proceedings of AAAI-2000*.

VII

HUMAN/ROBOT INTERACTION

HUMAN-ROBOT INTERACTIONS: CREATING SYNERGISTIC CYBER FORCES

Jean C. Scholtz
National Institute of Standards and Technology
100 Bureau Drive, MS 8940
Gaithersburg, MD.
jean.scholtz@nist.gov

Abstract   Human-robot interaction for mobile robots is still in its infancy. As robots increase in capabilities and are able to perform more tasks in an autonomous manner we need to think about the interactions that humans will have with robots and what software architecture and user interface designs can accommodate the human-in -the loop. This paper outlines a theory of human-robot interaction and proposes information needed for maintaining the user's situational awareness.

Keywords: Human-robot interaction, situational awareness, human-computer interaction

1.   INTRODUCTION

The goal in synergistic cyber forces is to create teams of humans and robots that are efficient and effective and take advantage of the skills of each team member. A subgoal is to increase the number of robotic platforms that can be handled by one individual. In order to accomplish this goal we need to examine the types of interactions that will be needed between humans and robots, the information that humans and robots need to have desirable interchanges, and to develop the software architectures and interaction architectures to accommodate these needs.

Human-robot interaction is fundamentally different from typical human-computer interaction in several dimensions. First, I propose that there are three different levels of interaction possible from the human view point, each with different tasks and interactions. I call these levels supervisory interactions, peer interactions, and mechanic interactions. The second dimension is the physical nature of mobile robots. Robots need some awareness of the physical world in which they move and this model needs to be conveyed to the human to facilitate an understand of the decisions made by the robotic platform. A third dimen-

A.C. Schultz and L.E. Parker (eds.), Multi-Robot Systems: From Swarms to Intelligent Automata, 177-184.

sion is the dynamic nature of the robot. Robots have physical sensors that may fail or degrade, affecting some of the platform's functionality, unlike the more static computer platforms. The final dimension is the environment in which interactions occur. Interaction devices and displays may have to function in harsh conditions and will have to support user mobility.

2.   SITUATIONAL AWARENESS

Situational Awareness has long been a metric for evaluating supervisory control interfaces and should be a good approach to developing metrics for HRI. However, each of the three levels of interaction will require a different set of information needs and hence different situational awareness.

Situational awareness (Endsley, 2000a) is the knowledge of what is going on around you. The implication in this definition is that you understand what information is important to attend to in order to acquire situational awareness. As you drive home in the evening there is much information you could attend to. You most likely do not notice if someone has painted their house a new color but you definitely notice if a car parked in front of that house starts to pull out in your path.

There are three levels of situational awareness (Endsley, 2000). Level One of situational awareness is the basic perception of cues. Failures to perceive information can result due to short comings of a system or they can be due to a user's cognitive failures. In studies of situational awareness in pilots, 76% of SA (Jones and Endsley, 1996) errors were traced to problems in perception of needed information. Level Two of situation awareness is the ability to comprehend or to integrate multiple pieces of information and determine the relevance to the goals the user wants to achieve. A person achieves the third level of situational awareness if she is able to forecast future situation events and dynamics based on her perception and comprehension of the present situation.

Performance and situational awareness, while related, are not directly correlated. It is entirely possible for a person to have achieved level three situational awareness but not perform well. A person can fail to perform correctly due to poorly designed systems or due to cognitive failures. The most common way to measure situational awareness is by direct experimentation using queries (Endsley, 2000). The task is frozen, questions are asked to determine the user's situational assessment at the time, then the task is resumed. The Situation Awareness Global Assessment Technique (SAGAT) tool was developed as a measurement instrument for this methodology (Endsley, 1988). The SAGAT tool uses a goal-directed task analysis to construct a list of the situational awareness requirements for an entire domain or for particular goals and subgoals. Then it is necessary to construct the query in such a way that the operator's response is minimized, such as presenting choices for responses.

3. A THEORY OF HUMAN-ROBOT INTERACTION

3.1 Level of Interaction Scenarios

To illustrate the three different levels of interaction, here are two different scenarios.

3.1.1 Scenario 1: Military Operations in an Urban Terrain.

A number of soldiers and robots are approaching a small urban area. Their goal is to make sure that the area is secured - free of enemy forces. A team of heterogeneous robots will move in a coordinated fashion through congested areas to send back images to the soldiers. Robots will also be used to enter buildings that may be hiding places for enemy soldiers. A supervisor is overseeing the scouting mission from a remote location close to but not in the urban area. Ground troops are close behind the robots and individual robots are associated with a certain group of the soldiers. In each group one soldier is an expert in maintaining the robot (both physically and programmatically) associated with that group. The supervisor needs to know that all robots are doing their jobs and reassigns robots appropriately as the mission moves forward. If there is a problem the supervisor can either intervene or can alert the soldier mechanic.

The soldier functioning as the mechanic does what is necessary to get the robot back into an operational state. This could be as simple as identifying an image or as complicated as debugging hardware and software. The soldiers are also team members of the robot and may give the robot tasks to do such as obtaining another image at closer range. The robots will also encounter civilians in the urban environment. Some degree of social interaction will be necessary so that the civilians don't feel threatened by the robots.

3.1.2 Scenario 2: Elder Care.

An elder care facility has deployed a number of robots to help in watching and caring for its' residents. The supervisor oversees the robots who are distributed throughout the facility and makes sure that the robots are properly functioning and that residents are either being watched or cared for — either by a robot or by a human caregiver. A number of human caregivers are experts in robot maintenance and assist as needed depending on their duties at the time. The caregiver robots can perform routine tasks such as helping with feeding, handing out supplies to residents, and assisting residents to move between locations in the facility. Watcher robots monitor residents and have the capability to send back continual video feeds but also to alert the supervisor or a nearby human caregiver to an emergency situation. Robots interact with the residents as well as visitors to the facility who may not be aware of their capabilities.

3.2   Scenario Generalizations

What can we learn from these two scenarios? First of all, the boundary between the three levels of interactions is fuzzy. The supervisor can take the mechanic role if it is more efficient than a handoff to the designated mechanic. The team members can command the robots. Peer interactions occur with team member and with bystanders who have little or no idea of the capabilities of the robot. All of the interaction levels can occur at the same time by the same or different people. Is it necessary and feasible to design adaptive interfaces that adjust as the user's role changes? What types of interaction technologies and feedback mechanisms are needed to support these interactions? What information is essential for users in each role?

3.3   The Supervisory Level

We assume that the supervisory interface is done from a remote location. Our hypothesis is that the supervisor needs the following information:

- an overview of the situation.

- the mission or task plan.

- current capabilities of any robotic platform and any deviation from "normal".

- other interactions with robots, including group robot behaviors.

A corresponding HCI domain is that of complex monitoring devices. Complex monitoring devices were originally based on displays of physical devices (Vincente, Roth, and Muman, 2001). The original devices were just lights and switches that corresponded to a sensor or actuator and were displayed on physical panels. Computer based displays were unable to hold all the information on a single display and so users had to sequence through a series of displays. This produced a keyhole effect— the notion that a problem was most likely occurring on a display that wasn't currently being viewed.

Another issue in complex monitoring devices is that of having an indication of what "normal" is. This is also true in human-robot interactions where physical capabilities of the system change and the supervisor needs to know the "normal" status of the robot at any given time. Another issue is that single devices may not be the problem but rather relationships between existing devices. Displays should support not only problem driven monitoring but knowledge driven monitoring when the supervisor actively seeks out information based on the current situation or task.

We suggest that lessons learned in producing displays for monitoring complex systems can be used as a starting point for supervisory interfaces for HRI.

In addition, basic HRI research issues that are not addressed in complex systems include:

- what information is needed to give an overview of teams of robots?

- can a team world model be created and would it be useful?

- what (if any) views of individual robot world models are useful?

- how to give an awareness of other interaction roles occurring?

- what strategies and information needs are appropriate for handing off interventions to others? to others?

Situational awareness indicators will be developed based on a task-analysis of the supervisor's role in a number of scenarios (such as those described earlier in this paper). An initial hypothesis about possible indicators of situational awareness includes:

- which robots have other interactions going on.

- which robots are operating in a reduced capability.

- the type of task various robots are currently carrying out.

- the current status of the mission.

3.4   Peer to Peer Interactions

We make the assumption that these are face to face interactions. This is the most controversial type of interaction. Our use of the terms "peers" and "teammates" is not meant to suggest that humans and robots are equivalent but that each contributes skills to the team according to their ability. The ultimate control rests with the user—whether the team member or the supervisor. The issue is how the user (in this case, the peer) gets feedback from the robot concerning its understanding of the situation and actions being undertaken. In human-human teams this feedback occurs through communication and direct observation. Current research (Bruce, Nourbakhsh, and Simmons, 2001; Breazeal and Scassellati, 1999) looks at how robots should present information and feedback to users. Bruce et al. stress that regular people should be able to interpret the information that a robot is giving them and that robots have to behave in socially correct ways to interaction in a useful manner in society. Breazeal and Scassellati use perceptual inputs and classify these as social and non-social stimuli, using sets of behaviors to react to these stimuli.

Earlier work in service robots (Engelhardt and Edwards, 1992) looked at using command and control vocabularies for mobile, remote robots including natural language interfaces. They found that users needed to know what commands were possible at any one time. This will be challenging if we determine that it is not feasible to have a separate device that can be used as an interface to display additional status from the robots that would be difficult to display via robotic gestures.

We intend to investigate research results in mixed initiative spoken language systems as a basis for communicating an understanding of the robot to the user and vice versa. Our hypothesis about information that the user will need include:

- What other interactions are occurring.

- The current status of the robot.

- The robot's current world model.

- What actions the robot can currently carry out.

Other interesting challenges include the constraint on the distance between the robot and the team. We use other communication devices to operate human-human teams from a distance. What are the constraints and requirements for robot team members?

3.5   Mechanic Interaction

We make the assumption that this will be either a remote interaction or will occur in an environment in which any addition cognitive demands placed on the user are by the environment are light. We will also assume that the mechanic has an external device to use as an interface to the robot. The mechanic must be a skilled user, having knowledge of the robotic architecture and robotic programming. If the robot has teleoperation capabilities, the mechanic would be the user who would take over control. This is the most conventional role for HRI - however, one that has not been extremely well supported. Moreover, as the capabilities and roles of robots expand, this role has to be capable of supporting interaction in a more complex situation.

We hypothesize that the mechanic needs the following information:

- The robot's world model.

- The real-world situation (as much as can be obtained).

- The robot's plans.

- The current status of any robotic sensors.

- Other interactions currently occurring.

- Any other jobs that are currently vying for the mechanic's attention (assuming it is possible to service more than one robot).

- The effects of any adjustments on plans and other interactions.

- Mission overview and any timing constraints.

Murphy and Rogers (Murphy and Rogers, 1996) note three drawbacks to telesystems in general:

- The need for a high communication bandwidth for operator perception and intervention.

- Cognitive fatigue due to repetitive nature of tasks.

- Too much data and too many simultaneous activities to monitor.

Simulation and programming environments and computer games should be examined to determine the properties present in good examples of these that contribute to the user's situational awareness.

4.   SUMMARY

We propose that human-robot interactions are of three varieties, each needing different information and being used by different types of users. In our research we will develop a number of scenarios, do a task-based analysis of the three types of human-robot interactions suggested by each scenario. We will then develop both a baseline interface for each type of interaction and a situational assessment measurement tool. We propose to conduct a number of user experiments and make the results publicly available.

We have concentrated on the user and her information needs in this paper. However, to achieve a successful synergistic team, it will be necessary to furnish information about the user to the robot and to create a dialogue space for team communication. We will start by concentrating on the user aspects of the information but intent to expand our research to include capture and use of user information as well.

Acknowledgments

This work was supported by the DARPA MARS program, NIST contract K300.

References

Breazeal C. and Scassellati,B. (1999). A context-dependent attention system for a social robot, *1999 International Joint Conference in Artifical Intelligence*

Bruce, A., Nourbakhsh, I., and Simmons. R. (2001). The Role of Expressiveness and Attention in Human-Robot Interaction, *AAAI Fall Symposium*, Boston MA, October.

Endsley, M. R. (1988). Design and evaluation for situation awareness enhancement. In *Proceedings of the Human Factors Society 32nd Annual Meeting* (Vol. 1, 97–1010) Santa Monica, CA: Human Factors Society.

Endsley, M. R., (2000). Direct Measurement of situation Awareness: Validity and Use of SAGAT in (Eds.) Mica R. Endsley and Daniel J. Garland. *Situation Awarenss Analysis and Measurement*, Mahwah, NJ: Lawrence Erlbaum Associates.

Endsley, M. R. (2000a). Theoretical Underpinnings of Situation Awareness: A Critical Review, in (Eds) Mica R. Endsley and Daniel J. Garland. *Situation Awareness Analysis and Measurement*. Mahwah, NJ: Lawrence Erlbaum Associates.

Engelhardt, K. G. and Edwards, R.A. (1992). Human-robot integration for Service robots (315–346) in (Eds) Mansour Rahimi and Waldemar Karwowski, *Human Robot Interaction*, London: Taylor and Francis.

Jones, D.G. and Endsley, M. R. (1996). Sources of situation awareness errors in aviation. *Aviation, Space and Environmental Medicine*, 67(6), 507–512.

Murphy, R. and Rogers, E. (1996). Cooperative Assistance for Remote Robot Supervision, *Presence*, volume 5, number 2, Spring, 224–240.

Vincente, K., Roth,E. and Muman, R. (2001). How do operators monitor a complex, dynamic work domain? The impact of control room technology, *Journal of Human-Computer Studies*, 54, 831–856.

COMMUNICATING WITH TEAMS OF COOPERATIVE ROBOTS

D. Perzanowski, A.C. Schultz, W. Adams, M. Bugajska, E. Marsh, J. G. Trafton, D. Brock
Codes 5512, 5513, and 5515, Naval Research Laboratory, Washington, DC 20375

M. Skubic
University of Missouri-Columbia, Computer Engineering & Computer Science Department, Columbia, MO 65211

M. Abramson
ITT Industries, Alexandria, VA 22303

Abstract: We are designing and implementing a multi-modal interface to a team of dynamically autonomous robots. For this interface, we have elected to use natural language and gesture. Gestures can be either natural gestures perceived by a vision system installed on the robot, or they can be made by using a stylus on a Personal Digital Assistant. In this paper we describe the integrated modes of input and one of the theoretical constructs that we use to facilitate cooperation and collaboration among members of a team of robots. An integrated context and dialog processing component that incorporates knowledge of spatial relations enables cooperative activity between the multiple agents, both human and robotic.

Keywords: Cooperative and collaborative behavior, dynamic autonomy, human-robot interaction, multi-modal interfaces

A.C. Schultz and L.E. Parker (eds.), Multi-Robot Systems: From Swarms to Intelligent Automata, 185-193.

1.   INTRODUCTION

Interacting and communicating with another person is a complicated set of processes in real life. However, humans learn and master the linguistic and social skills necessary to perform this feat with seemingly little effort. In just a few short years they are capable of carrying on conversations and other interactions with another human and in most cases have little difficulty extending these skills to perform similar functions with a group of individuals. Certain aspects of communication enable individuals to form teams to achieve their goals. We are interested in investigating what those aspects of communication are, and then incorporate them in our multi-modal interface for human-robot interactions.

We have already incorporated natural language and gestures into our human-robot interface (Perzanowski, *et al.*, 1998, 2000); we are now introducing additional context and dialog processing to facilitate natural communication and enable cooperative action between multiple agents.

2.   COMMUNICATION ISSUES

2.1   Linguistic Cues

Certain contextual and linguistic cues provide crucial information to humans for them to communicate easily. Prosodic cues, such a the inflection of one's voice, the rise and lowering of pitch of the voice, tell the participants of a dialog that an utterance is being made, a certain type of utterance is being made, and that the utterance is ending, or more is to come. However, state-of-the-art speech recognition engines sensitive to this kind of information are not commercially available. It would seem that greater cooperation and teamwork between humans and robots is stymied by the inability of speech recognition engines to provide important information to participants in a dialog. However, other cues used by humans enable them to interact and exchange information during a dialog. We currently use the syntactic and semantic information that both our speech recognition system, ViaVoice, and natural language understanding system, Nautilus (Wauchope, 1994), provide. Additional contextual information is obtained from visual cues, spatial information and an analysis of the linguistic information available to us in *context predicate*s (Perzanowski, *et al.*, 1999) to foster collaboration and cooperation in a team of human and robot agents. We turn now to a discussion of these features.

2.2   Visual Cues

Visual cues, such as "body language," provide humans with the kinds of information needed to facilitate dialog and promote teamwork. For example, if the speaker of sentence (1) is standing in front of two individuals but staring at one of them, then it is incumbent upon the person being stared at to respond in some way.

(1) The computer is over there.

Likewise, the speaker of (1) might gesture—point—or simply shrug a shoulder in a particular direction to indicate information about the location of the object.

Finally, participants in a dialog may either directly address whom they wish to perform certain actions, as in (2), or they may focus their attention on a person or a thing.

(2) Coyote, go to the computer on the left side of the room.

Eye gaze directed at Coyote, without directly addressing Coyote in (2), cues all the listeners of the utterance to the fact that the speaker wishes Coyote to perform the action. Nodding one's head at the listener can indicate the same intentions. Therefore, visual cues can be utilized by an interface to compensate for the lack of certain information.

2.3   Knowledge

Knowledge of the various participants and the environment can also facilitate collaborative communication. For example, if someone knows that a person can only make group meetings on Fridays at 10 o'clock, a great deal of extraneous communication can be avoided, given such a pre-condition. Likewise, knowledge of the capabilities—the strengths and/or weaknesses—of the various agents in a dialog can benefit communication.

Asking someone to lift an object when that person is not capable of doing so is counter-productive. Likewise, if one of the sensors on a robot team member suddenly fails and is no longer usable, sharing this information with the other participants can prevent extraneous communication and wasting time.

Environmental information, such as spatial knowledge (Skubic, et al. 2002), can also assist team members in achieving their goals. Determining that an object is within range of the sensors of one robot, and having that robot communicate this information to the other participants, contributes to a more timely solution to the task.

In our initial research, we focused on natural language and natural gestures in command and control situations with a single robot or multiple robots that still acted independently. We are working with a team of dynamically autonomous robots interacting without constant human intervention. We define the term "dynamically autonomous" to mean that the agents are capable of operating at varying levels of autonomy, based on their individual awareness of their own capabilities in achieving some goal; their awareness of other agents' capabilities; and their knowledge of the overall plan (Pollack and McCarthy, 1999) and the history of achieving subgoals as the overall plan progresses (Grosz, et al. 1999).

3.   MULTI-MODAL INTERFACE

3.1   Gesture and Object Recognition

We are using several robots, Nomad 200s, XR-4000s, and an RWI ATRV-Jr. Gestures are detected using a structured light rangefinder. A camera fitted with a filter tuned to the laser wavelength is mounted on its side. The robot is capable of tracking the user's hands and interpreting their motion as vectors or measured distances. A more detailed discussion can be found in (Perzanowski, *et al.*, 1998). Sonar sensors on the robots detect objects in the environment. With this data, object recognition is possible (Skubic, *et al.*, 2001a, 2001b). We are currently incorporating a binocular vision system to permit more sophisticated recognition of both objects and people.

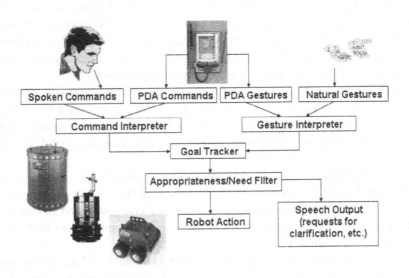

Figure 1. Multi-modal Interface

The interface (Figure 1) also employs a PDA with a stylus and touch-screen. Pointing, clicking or drawing on the touch screen indicate locations, regions, directions, and the like.

3.2   Natural Language Processing

A more detailed description of the natural language processing is discussed elsewhere (Perzanowski et al. 1998), but a brief discussion here introduces one element of the dialog which we employ for collaboration and cooperation in achieving goals. Vocal commands or clicks on buttons on the PDA screen are mapped into a logical form. The latter is correlated with gesture data, knowledge of the other participating agents, and with spatial information obtained from the robot sensors. The result is then mapped to a robot command, which produces either some action or an interchange of information. For example, the human user can direct a robot by uttering sentence (3).

(3) Coyote, go to the north side of the nearest building.

The spatial reasoning component uses the sensor data to determine that an object exists and it computes where the north side of the object is. If the sensors detect an appropriate object, the various inputs are combined and a robot command is sent to the robot to act accordingly. If, on the other hand, no such object is sensed, the robot complains verbally, saying something to the effect that no such object exists.

We track information about goals, i.e. whether or not goals have been attained, in *context predicates*. Context predicates are linguistically motivated constructs that contain semantic and contextual information of the discourse. (4) is the context predicate for (3).

         (4) ((imper (:verb gesture-go
                 (:agent (:system you))
                 (:to-loc ((:thing side)
                      (:dir north))
                   ((:relation-to building)
                   ((:descrip nearest)
                     (:relation-to you))))) 1).

If a goal is achieved, the context predicate reflects this, as signified by the "1" in the representation. If the goal is not achieved, the representation exhibits a "0." As the discourse continues, the stack of context predicates is updated: if the focus of the dialog changes, completed goals are eliminated, but non-completed goals remain. Since this knowledge is shared by all of the participants in the dialog, anyone can act upon the non-completed goals, if the situation warrants it. Thus, if for some reason Coyote is unable to complete its specified goals, another robot can be tasked to complete the goals.

As the dialog progresses, the focus of the dialog changes (Grosz and Sidner, 1986). Keeping track and updating the focus of the dialog updates the context predicates.

We are currently interested in having robots determine on their own-- based upon a particular task, their individual capabilities, knowledge and overall plan (Grosz, et al. 1999)--what teams should be formed, and who is a member of which team. Tasks can be achieved with as little human intervention as possible. Once the initial task is given, robots can form their own groups and obtain the goals more easily because they group themselves according to their individual strengths and appropriateness for completing certain goals. Thus, for example, armed robots would determine that they would be the best candidates for certain kinds of operations, while robots not so equipped would be more appropriate candidates for other missions. Furthermore, if one robot is tasked to go to a building, but another is closer, we are building in the capability to permit the latter robot to intervene and perform the action.

4.   RELATED WORK

We are attempting to incorporate linguistic and visual information into a multi-media interface to foster collaborative and cooperative teamwork. Other models incorporating collaboration and discourse theory exist, such as COLLAGEN (Rich, et al. 2001) and TRIPS (Allen, et al. 2001). Like COLLAGEN, we are grounding our work in linguistic and discourse theory and attempting to make the interface application-independent. However, we incorporate context predicates from the discourse, and unlike COLLAGEN we are using visual cues and spatial information to motivate team formation and teamwork.

TRIPS already incorporates much of the collaborative kinds of interaction we are looking for in a dialog. However, with our emphasis on context predicates, we are hoping to minimize human intervention in the collaboration.

Our emphasis on multi-modal and natural interaction sets us somewhat apart from the work of (Fong, et al. 2001). This research does not emphasize natural language in their interface to control a robot, and natural gestures are not employed. Instead, their interactions are limited to a set of messages and their gesturing is viewed as a translation of gestures into a visual joystick.

We, on the other hand, are interested in natural commands and visual interactions with robotic agents. While our work incorporating a PDA device is very similar, we have not attempted any interface with a Web-based interface at this time. However, our goal is identical: development of a system in which humans and robots work together as cooperative agents in performing some task.

5. FUTURE WORK

While we do not incorporate a Web-based interface presently, we are working on adding this capability. In the future, we hope to access online information about novel locations, so that the robots can navigate through unknown terrain, having obtained information about routes and the environment from internet sources.

We are currently expanding our knowledge component to incorporate vocabulary acquisition in real-time. At present, if an object is sensed, and the human user tells a robot that the object is called a "computer," for example, the spatial reasoning component maintains this information, but it is not passed to the natural language understanding component. In other words, while the object "computer" exists in a robot's sensor readings and in its knowledge of the space around it, it still cannot communicate information about the computer naturally. Simply, while it knows that a computer exists, it cannot talk about it, or perform some rather rudimentary reasoning about the object so labeled.

We are, therefore, working on adding the ability to reason about objects. Thus, if an object is perceived from a certain viewpoint, we are adding the ability to know that an object, let's say a computer, is the same computer if viewed from a different point of view. We would also like for our team of robots to know that objects once identified, if moved, are still the same objects. Only their locations have changed.

We continue to focus our attention on the use of context predicates and a dialog-based planning component to motivate team formation and teamwork

6. CONCLUSION

We are concentrating on two main research areas to facilitate cooperation and collaboration in a team of robots. The first area focuses on context predicates, linguistically motivated constructs that contain semantic and goal information. Using context predicates, teams of robots share information about goal status and act accordingly.   The second research area is our expansion of the spatial reasoning component so that robots reason about their physical environment and share information about the environment, objects, and locations.

Our purpose is to enhance team formation and dynamic autonomy so that robots interact with each other and human intervention occurs only as needed.

Acknowledgments

This work was partially funded by the Office of Naval Research and the DARPA ITO MARS program.

References

Allen, J., Byron, D.K., Dzikovska, M., Ferguson, G., Galescu, L., and Stent, A.   (2001). Toward Conversational Human-Computer Interaction. *AI Magazine,* (22)4:27-37.

Fong, T., Thorpe, C., and Baur, C. (2001). Advance Interfaces for Vehicle Teleoperation: Collaborative Control, Sensor Fusion Displays, and Remote Driving Tools. *Autonomous Robots,* 11: 77-85.

Grosz, B. and Sidner, C. (1986).   Attention, Intentions, and the Structure of Discourse. *Computational Linguistics*, 12(3):175-204.

Grosz, B., Hunsberger, L. and Kraus, S. (1999). Planning and Acting Together. *AI Magazine*, 20(4): 23-34.

Perzanowski, D., Schultz, A.C. and Adams, W. (1998).   Integrating Natural Language and Gesture in a Robotics Domain. In *Proc. IEEE Int'l Symp. Intelligent Control*, pages 247-252, Piscataway, NJ.

Perzanowski, D., Schultz, A., Adams, W., and Marsh, E. (1999). Goal Tracking in a Natural Language Interface: Towards Achieving Adjustable Autonomy. In *Proc. 1999 IEEE Int'l Symp. Computational Intelligence in Robotics and Automation*, pages 144-149, Piscataway, NJ.

Perzanowski, D., Adams, W., Schultz, A., and Marsh, E. (2000). Towards Seamless Integration in a Multimodal Interface. In *Proc. 2000 Workshop Interactive Robotics and Entertainment*, pages 3-9, Menlo Park, CA.

Pollack, M. and McCarthy, C. (1999). Towards Focused Plan Monitoring: A Technique and an Application to Mobile Robots. In *Proc. 1999 IEEE Int'l Symp. Computational Intelligence in Robotics and Automation*, pages 144-149, Piscataway, NJ.

Skubic, M., Perzanowski, D., Schultz, A., and Adams, W. (2002). Using Spatial Language in a Human-Robot Dialog. In *2002 IEEE Int'l Conf. on Robotics and Automation.*

Skubic, M., Chronis, G., Matasakis, P., and Keller, J. (2001a). Generating Linguistic Spatial Descriptions from Sonar Readings Using the Histogram of Forces. In *Proc. of the 2001 IEEE Int'l Conf. on Robotics and Automation*, Seoul, Korea.

Skubic, M., Chronis, G., Matasakis, P, and Keller, J. (2001b). Spatial Relations for Tactical Robot Navigation. In *Proc. of the SPIE, Unmanned Ground Vehicle Technology III*, Orlando, FL.

Rich, C., Sidner, C., and Lesh, N. (2001). COLLAGEN: Applying Collaborative Discourse Theory to Human-Computer Interaction. *AI Magazine,* 22(4):15-25.

Wauchope, K. (1994). *Eucalyptus: Integrating Natural Language Input with a Graphical User Interface*, Technical Report NRL/FR/5510-94-9711, Naval Research Laboratory, Washington, D.C.

ROBOT AS PARTNER: VEHICLE TELEOPERATION WITH COLLABORATIVE CONTROL

Terrence Fong and Charles Thorpe
The Robotics Institute, Carnegie Mellon University, Pittsburgh, Pennsylvania

Charles Baur
Swiss Federal Institute of Technology, Lausanne, Switzerland

Abstract   We have developed a new teleoperation system model called *collaborative control*. With this model, the robot asks the human questions, to obtain assistance with cognition and perception during task execution. This enables the human to support the robot and to compensate for inadequacies in autonomy. In the following, we review the system models conventionally used in teleoperation, describe collaborative control, and discuss its use.

Keywords: Collaborative control, human-robot interaction, vehicle teleoperation.

1.   INTRODUCTION

In teleoperation, a robot is commonly viewed as a tool: a device capable of performing tasks on command. As such, a robot has limited freedom to act and will always perform poorly whenever its capabilities are ill-suited for the task at hand. This is particularly true when high-level perceptual functions (e.g., object recognition) are involved. Moreover, even if a robot realizes that it is performing poorly, it usually has no way to ask for (or to gain) assistance.

The problem is that the "robot as tool" paradigm is extremely limiting, i.e., it restricts the human-robot relationship to that of master-slave. As a result, the system's capability is strictly bound to the operator's skill and the quality of the user interface. In order to make teleoperation better performing, therefore, we must find a new approach. What we need is a paradigm that is more flexible, that encourages human-robot synergy, and that allows robots to work as partners (if not as peers).

A.C. Schultz and L.E. Parker (eds.), Multi-Robot Systems: From Swarms to Intelligent Automata, 195-202.
© 2002 *Kluwer Academic Publishers. Printed in the Netherlands.*

2.   COLLABORATIVE CONTROL

2.1   A Robot-Centric System Model

To address this need, we have developed *collaborative control*, a system model in which human and robot work together (Fong, 2001; Fong, 2002). Instead of a supervisor dictating to a subordinate, the human and the robot engage in dialogue to exchange information, to ask questions, and to resolve differences. With this approach, human-robot interaction is more natural, more balanced, and more direct than conventional approaches.

With collaborative control, the human functions as a resource for the robot, providing information and processing just like other system modules. In particular, the robot is allowed to ask the human questions as it works, to obtain assistance with perception and cognition. This allows the human to compensate for limitations of autonomy. Moreover, since the robot is aware that the human may not respond, collaborative control enables dynamic, fine-grained sharing of control. With other forms of teleoperation, the division of labor is pre-defined or is coarsely switched on a per-task basis.

To understand how this works in practice, consider the following situation: a mobile robot is driving forward when it has difficulty deciding if there is an obstacle in its way (e.g., range sensors return conflicting information). At this point, the robot must make a decision. With conventional design, there are three choices: wait for the path to become clear, look for a way around, or ignore the conflict and continue forward. All of these strategies have significant problems: "wait until clear" may cause indefinite delay; "make a detour" may consume excessive resources; and "drive through" may result in excessive damage.

With collaborative control, the robot can ask for human assistance. In this case, the robot can present the sensor data (e.g., a camera image) to the human and ask his opinion. Once the human sees the data, he may decide (based on experience or interpretation) that "drive through" is acceptable. In other words, through collaboration, the robot can avoid needless delay and having to make an unnecessary detour.

Collaborative control provides an effective framework for coordination. Since the human can only attend to one module (or robot) at a time, we arbitrate among the requests to select which is presented. This allows human attention to be directed where it is most needed, in terms of safety, priority, etc. Additionally, because collaborative control incorporates a user model for dialogue management, it can accommodate users with varied backgrounds and capabilities. Thus, the robot can ask questions and can interpret responses, based on the user's expertise, preferences, and other characteristics.

3. RELATED RESEARCH

Some robot control architectures have addressed the problem of mixing humans with robots. One method is to incorporate humans as a system module (Rosenblatt, 1995). Another method is prioritized control, the classic example of which is NASREM (Albus, 1987). In both these approaches, the human only provides command input. With collaborative control, however, the human contributes his expertise wherever it can be used.

Teleassistance tries to improve teleoperation by supplying aid to the operator in the same manner that an expert would (Murphy and Rogers, 1996). Collaborative control takes the opposite approach: it provides the robot with human assistance. Although adjustable autonomy (e.g., (Dorais, 1999)) shares some aspects of collaborative control, human-robot dialogue is not used as a mechanism for adaptation and coordination.

In many ways, collaborative control is most similar to remote expert systems. For example, Bauer et al. describe a wearable computer with a camera that enables office-based experts to see what a field technician sees (Bauer, 1998). The parallel with collaborative control is clear: the robot is like the field technician (i.e., it is skilled, but may need help) and the operator is like the expert (i.e., he can provide assistance when needed).

4. CONVENTIONAL SYSTEM MODELS

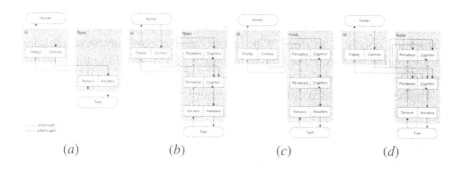

Figure 1. Teleoperation system models: a, direct control; b, supervisory control; c, fully autonomous control; d, collaborative control.

The most common method for vehicle teleoperation is *direct control*: the operator operates the vehicle using hand-controllers while monitoring video displays (Figure 1a). Because all control decisions depend on the human, system performance is directly linked to human capabilities. Many factors including skill, training, etc., all play a role in how the system functions. Other factors, such as communication bandwidth, may also influence operational efficacy.

The *supervisory control* system model is shown in Figure 1b. With supervisory control, the human divides a problem into a sequence of tasks which the robot performs on its own (Sheridan, 1992). Once he gives control to the robot, the human typically assumes a monitoring role. However, the human may also intermittently (i.e., trade or share) control the robot by closing a command loop or he may control some variables while leaving the others to the robot.

Fully autonomous control is somewhat of a misnomer because it rarely is fully automatic. With this system model, the human gives high-level goals, which the robot independently achieves (Figure 1c). The difference between supervisory and fully autonomous control is the nature of the goal. In the former, goals are limited and task planning is performed primarily by the human. With the latter, goals are more abstract and the robot is responsible for planning.

With conventional system models, poor performance (or failure) will occur if the human fails to recognize that the robot is ill-suited for the task or situation. Additionally, none of the models can effectively accommodate a wide range of users. Direct control, for example, is generally limited to trained, expert users because difficulty and risk are high.

5.   COLLABORATIVE CONTROL SYSTEM MODEL

Collaborative control addresses the limitations of conventional system models through collaboration and dialogue. In supervisory or fully autonomous control, if the robot has difficulties, the only choices it has are to continue performing poorly or to stop. With collaborative control, however, the robot has the option of asking the human to assist: providing information, helping perform perception or cognition, etc.

Another way in which collaborative control differs from conventional system models is that it provides fine-grained sharing/trading of control and autonomy adjustments. Because work is dynamically allocated through dialogue, the human is automatically included in the control loop as needed. This is a significant difference from other models, which require the user to decide how, when, and where control should be allocated.

With collaborative control, therefore, the human may be involved in multiple control loops (Figure 1d). As with supervisory control, he may close a command loop or monitor task execution through interface displays. As a resource for the robot, however, the human may also close a perception loop, a cognition loop, or some combination of the two. Furthermore, the human may interact with the robot at different levels of abstraction.

6. SYSTEM DESIGN

We have implemented collaborative control as a set of modules (dialogue management, robot control, etc.), each of which operates with variable autonomy in a message-based architecture (Fong, 2001). Whether a module operates at a low level (strongly dependent on the human) or high level (little or no human interaction) of autonomy is determined by factors including situational demands, module competency, and user model.

In our system, dialogue is the exchange of messages between human and robot. Our current system has approximately thirty messages: human-to-robot (command, queries, responses) and robot-to-human messages (information statements, queries). At present, the robot is able to ask two types of queries. *Safeguard* queries concern safety issues (e.g., "Stopped due to rollover danger. Can you come over and help?". *Task* queries describe task-specific functions, such as "Motion detected. Is this an intruder (*image*)? If you answer *yes*, I will follow him."

Each robot query is described by a number of attributes, some of which are operator-dependent (required response accuracy, required expertise) and others which are operator-independent (expiration, priority, etc). These attributes are used to select which queries will be asked. A failure to respond, whether intentional or not, can trigger behavioral changes (e.g., the robot's level of autonomy may be increased).

7. RESULTS

7.1 Remote Driving Tests

During the past year, we have conducted a variety of remote driving tests to evaluate the use of collaborative control (Fong, 2001; Fong, 2002). In one test, we examined the use of multiple mobile robots for reconnaissance and surveillance. One of the attractive features of collaborative control is that it directs the human's limited resources (attention, cognition, etc.) where they are needed. This relieves the human of the burden of simultaneously monitoring each robot.

Figure 2 shows an experiment in which an operator used two robots to perform outdoor reconnaissance. During the test, each robot asked safety-related questions while traversing unknown terrain. For example, both robots had questions about setting safety levels. Although it is common practice to define "normal" safety levels, there are some occasions when a system must be used beyond its design specifications. This is particularly true for situations in which system loss is acceptable as long as the goal is achieved (e.g., military combat missions).

Figure 2.   Outdoor reconnaissance with two robots

7.2   User Study

To examine how collaborative control influences human-robot interaction, we recently performed a Contextual Inquiry (CI) user study (Fong, 2001). CI is a structured interviewing method for grounding interactive system design in the context of the work (Holtzblatt and Jones, 1993). In the study, users were required to explore a cluttered environment while assisting safeguarding autonomy.

The study revealed several interesting findings. We observed that: (1) different users may respond quite differently to the same question; (2) users may grow weary of answering questions; (3) a question without adequate detail is hard to answer; (4) dialogue can make users personify the robot; and (5) indicating the urgency of questions is important.

Overall, we found dialogue to be valuable for teleoperation. In particular, novices reported that dialogue significantly helped them understand the problems the robot encountered during task execution. Although experts were generally less satisfied than novices, primarily because they grew tired of answering questions, they also stated that dialogue was a useful in keeping them involved and engaged in system operation.

8. DISCUSSION

8.1 Benefits of Collaborative Control

Unlike other forms of teleoperation, in which the division of labor is defined *a priori*, collaborative control allows human-robot interaction and autonomy to vary as needed. If the robot is capable of handling a task autonomously, it can do so. But, if it cannot, the human can provide assistance.

The use of dialogue makes human-robot interaction adaptable. Since the robot is aware of the user, it can always decide if asking a question will be useful. Because it has knowledge of the human's expertise, accuracy, etc., the robot can consider whether to accept a response at face value or to weigh it against other factors. In this way, system operation can adapt to different operators.

Dialogue helps the human to be effective. By focusing attention where it is needed, dialogue helps coordinate and direct problem solving. In particular, in situations in which the robot does not know what to do, or when it is working poorly, a simple human answer (even a single bit of information from a novice) is often all that is required to get the robot out of trouble.

8.2 Limitations of Collaborative Control

Although collaborative control is beneficial to teleoperation, there are limits to what it can provide. If human-robot interaction is adaptive, then control and information flow will vary with time and situation. This can make validation and verification difficult because it becomes harder to duplicate an error condition or a failure situation.

Another consideration is that when humans and robots interact to achieve common goals, they are subject to team related issues. In particular, teamwork requires team members to coordinate and synchronize their activities, to exchange information and communicate effectively, and to minimize the potential for interference between themselves. Moreover, there are numerous factors which can impact and limit group performance including resource distribution, timing, progress monitoring, and procedure maintenance.

Finally, working in collaboration requires that each partner trust and understand the other. To do this, each collaborator needs to have an accurate model of (1) what the other is capable of doing and (2) how he will carry out a given assignment. If the model is inaccurate, or if the partner cannot be expected to perform correctly, then the collaboration will not work well.

9.   OPEN ISSUES

As we have discussed, collaborative control provides a framework for co-ordinating and adapting robot operation. The issue of scalability, however, remains to be addressed. For example, if the human must interact with a large number of robots, it might not be possible for him to assist each one individually. Instead, it would be more practical to focus on group interaction and have the robots work in formation. Similarly, if a robot has many varied modules, it may be difficult for the human to help different types of autonomy. In this case, it might be more efficient for the human to only assist certain modules.

In more general terms, human assistance is clearly a limited resource. Hence, we need to find ways of motivating the user to respond. This is particularly important when the robot must operate for long periods. One approach would be to develop an algorithm for choosing questions that are "significant" to the user, i.e., having some level of information theoretical content (bits), matching user interest, etc.

Acknowledgments

This work was partially supported by grants from the DARPA ITO MARS program, the National Science Foundation, and SAIC.

References

Albus, J., et al. (1987). *NASA/NBS Standard reference model for telerobot control system architecture (NASREM)*. Technical Note 1235, NIST.

Bauer, M., et al. (1998). A Collaborative Wearable System with Remote Sensing. In *Proc. of 2nd International Symposium on Wearable Computers*.

Dorais, G., et al. (1999). Adjustable autonomy for human-centered autonomous systems. In *Proc. of Workshop on Adjustable Autonomy Systems (IJCAI)*.

Fong, T. (2001). *Collaborative control: a robot-centric model for vehicle teleoperation*. PhD thesis, Carnegie Mellon University.

Fong, T., et al. (2002). Multi-robot remote driving with collaborative control. *IEEE Transactions on Industrial Electronics* (in press).

Holtzblatt, K. and Jones, S. (1993). Contextual inquiry: a participatory technique for system design. In D. Schuler and A. Namioka (editors), *Participatory design: principles and practice*, Lawrence Erlbaum.

Murphy, R. and Rogers, E. (1996). Cooperative assistance for remote robot supervision. *Presence*, 5(2).

Rosenblatt, J. (1995). DAMN: A distributed architecture for mobile navigation. In *Proc. of AAAI Spring Symposium on Lessons Learned from Implemented Software Architecture for Physical Agents*.

Sheridan, T. (1992). *Telerobotics, automation, and human sup. control*. MIT Press.

ADAPTIVE MULTI-ROBOT, MULTI-OPERATOR WORK SYSTEMS

Aaron C. Morris, Charles K. Smart, and Scott M. Thayer
Robotics Institute - Carnegie Mellon University
5000 Forbes Avenue
Pittsburgh, PA 15213
{acmorr, cks, sthayer}@cs.cmu.edu

Abstract: Unstructured and hostile environments impose great risk to exposed humans and present ideal domains for robotic forces; however, these dynamic environments pose considerable difficulty in autonomous multi-robot coordination, making a need for supervisory control paramount. This paper examines a three-phase approach that increases the robustness, reliability, and efficiency of human-machine work systems by dynamically altering the soldier-robot control relationships as well as the effective autonomy manifested by each robot function in response to estimated cognitive loading (stress). This approach enables an adaptive command and control structure across a spectrum of force configurations.

Keywords: Human-machine work systems, command and control, supervisory control

1. INTRODUCTION

Unstructured and hostile environments impose great risk to exposed humans and present ideal domains for robotic applications; however, the conditions in such environments represent complex challenges for autonomous robot coordination. The uncertainties that exist in dynamic environments (i.e., dynamic obstacles, insufficient prior knowledge, etc.) can impede the attainment of robot objectives thereby reducing the dependability of such systems for military operations. Enhancements to autonomous multi-robot coordination are required for reliable performance in unstructured

A.C. Schultz and L.E. Parker (eds.), Multi-Robot Systems: From Swarms to Intelligent Automata, 203-211.
© 2002 *Kluwer Academic Publishers. Printed in the Netherlands.*

domains. The approach proposed in this paper utilizes human cognition and dynamically reconfigurable control structures to enhance the reliability of robot work systems.

The integration of human and machine has long been investigated and found to be mutually advantageous (Jordan, 1963). Recent research, (Montemerlo, 2000) and (Kortenkamp, *et al.*, 1991), has developed adjustable autonomy to allow proportional alterations in robot autonomy so that humans can opportunistically inject guidance into autonomous command and control loops. Additionally, cognitive science, (Lebiere, 2001), has demonstrated successful modeling of human cognition so that situations of cognitive overload can be predicted and corrected before problems are manifested.

This paper outlines the merger of computational cognitive estimation with robot control theory to create a system that permits a small group of human operators the ability to manage a fleet of semi-autonomous robots within a spectrum of operational scenarios. This work is presented as a three-phase system (workload estimation, workload optimization, and workload mitigation) detailing the concept, mathematical background, and algorithm that form the linkage between human cognition and robot control.

2.   THE SYSTEM OVERVIEW

An operator-robot work system embodies a set of tasks (**T**) assigned to a force of heterogeneous robotic units (**R**) that are managed by a small crew of human operators (**H**). Tasks represent a wide range of duties (e.g. moving from one point to another, mapping, etc.) that robotic units are expected to perform. Humans represent a finite resource of cognition to aid and assist robot action. Robots are represented as sets of capabilities, or **subsystems** (**S**), and the atomic actions, or **functions** (**F**), that compose these subsystems. Modeling robots at a functional level allows (1) operators to enact control to specific actions providing a more efficient distribution of robot resources and (2) adaptive robot management since operator attention can be tuned upon to an optimized set of actions (Figure 1).

Figure 1. Decomposition of robot into subsystems and functions

The mathematical representation of these components is $H = \{h_1, h_2, ..., h_M\}$, $R = \{r_1, r_2, ..., r_N\}$, and $T = \{t_1, t_2, ..., t_P\}$ for M operators, N robots, and P tasks, respectively. Furthermore, a robot is a set of subsystems (or a subset of all subsystems) $r = \{s_1, s_2, ..., s_i, ..., s_K\} \Rightarrow r \subseteq S$ and a subsystem is a set of functions (or a subset of all functions) $s_i = \{f_1, f_2, ..., f_j, ..., f_J\} \Rightarrow s_i \subseteq F$.

System component interaction (i.e. the interaction between humans and robots) plays an important role in the configuration of the system. To complete the system definition, the following list of axioms detail a base set of relationships that can exist among system components.

1. A function can be controlled by only one operator
2. An operator can control multiple functions
3. One task can occupy multiple subsystems
4. A single subsystem can concurrently execute multiple tasks

3. SYSTEM COMPONENTS

3.1 Workload Estimation

As stated, operator cognition serves to supplement autonomous control when intrinsic decision-making capabilities hinder expected robot action; however, human cognition is an exhaustible resource, which limits the scope of operator control (Lebiere, 2001). Using the ACT-R cognitive architecture (Lebiere, 2001), human cognition can be modeled and estimated as a means of regulating operator control activity to prevent cognitive overloading. According to (Anderson and Lebiere, 1998), cognition can be separated into six categories (e.g. working memory, long-term memory, vision, motor, speech, and audition.) that compose cognitive processes. Quantifying these cognitive areas produces a **cognitive capacity vector**

($\vec{\Lambda} = [\lambda_{wm}, \lambda_{ltm}, \lambda_v, \lambda_s, \lambda_m, \lambda_a]^T$) that numerically describes the current/future cognitive status of an operator given the environmental feedback.

This formal representation of human cognition can be extended to represent the cognitive load induced from managing robot functions. The **cognitive loading vector** ($\vec{\Gamma} = [\varphi_{wm}, \varphi_{ltm}, \varphi_v, \varphi_s, \varphi_m, \varphi_a]^T$) describes the cognitive capacity consumed in each cognitive area through function interaction. It is important to note that varying the level of function autonomy will alter the amount of cognitive capacity consumed; however, this matter will be addressed in the following sections.

3.2   Workload Mitigation

To enable robot functions the ability to adapt to environmental dynamics, the autonomy level controlling function behavior must also be dynamic. Adjustable autonomy et al (Montemerlo, 2000; Kortenkamp, *et al.*, 2000; and Fong, *et al.*, 1999) provides a variable level of human-function interaction that mitigates the workload of operators and robots to the requirements of the system.   Derived from (Korte and Vygen, 1991), the four levels of function autonomy used in this work are:

1. **Fully autonomous**: periodic monitoring of function progress with infrequent, asynchronous operator interaction.
2. **Semi-autonomous**: minimal operator intervention with infrequent, yet sometimes critical, levels of operator interaction
3. **Indirect Manual Control**: discrete and periodic intervention with moderate to relatively high levels of operator interaction
4. **Manual Control**: continuous and direct intervention with consistently high (and perhaps maximal) levels of operator interaction

The cognitive load induced from a fully autonomous function (requiring little attention) will be less demanding than a manually controlled function (requiring constant attention) identifying a relationship between cognitive loading and functional autonomy. More formally, this relationship varies induced cognitive load as functional autonomy transitions from one level to another. To model the effects of autonomy transitions on cognitive loading, a set of linear transforms (Δ_k for k = 1...4 autonomy levels) are established. Mathematically, this representation permits the occurrence of level-to-level autonomy transitions to dynamically recalculate the cognitive loads of the affiliated robot function.

3.3   Workload Optimization

As defined, operator-robot work systems can become extremely complex and intractable structures since multiple tasks are distributed over multiple functions and managed by multiple operators. Even then, dynamic adjustment of autonomy exponentially inflates the number of operator-function combinations creating a vast configuration space. The required solution will involve searching this configuration space to determine a preferable set of operator-functions matches.

To develop the search technique, the definitions from the previous sections are condensed and composed into an equation of matrices. Combining all cognitive capacity vectors into a **capacity matrix (H_C)** and all cognitive loading vectors into a **loading matrix (R_C)** the actual cognitive load of the system (H_L) is determined by $R_C M = H_L$ where M is Boolean configuration matrix. The role of workload optimization is to select an M that meets the following criteria:

1. H_L is minimized or at most $H_L < H_C$ for all matrix elements enforcing the cognitive load to be bounded by human cognitive capacity.
2. The number of managed functions (i.e. the number of rows in M that contain a non-zero element) is maximized enforcing that no functions are left unsupervised during task completion.

These constraints work in opposition to one another: minimization seeks to preserve cognitive capacity while maximization attempts to consume it. In situations where both conditions cannot be satisfied, a compromise must be established so that the system is optimally arranged to operate with maximum efficiency. This paper represents work efficiency as a system cost function that processes several system-critical parameters. These parameters include

1. **Functional Priority**: the relative importance of a function (e.g. high-priority functions should be given precedence when cognition is limited)
2. **Preferred Configuration**: the default configuration of operators and functions (e.g. the system should preserve this configuration whenever possible)
3. **Authorization**: the permission granting or denying an operator-function match (e.g. unauthorized matches should not exist)
4. **System Anticipation**: the elapsed time of a function/operator request for autonomy alteration (e.g. prevent cognitive starvations)

5. **Task Inertia**: the elapsed time between a linked operator and function (e.g. prevent unnecessary context switching among assignments)

By quantifying these parameters, the cost function can be tailored to suit the requirements of any work system and allow efficiency to be determined by comparing relative system cost. For example, the minimum system cost can be cast as the goal for searching the configuration space. The resulting configuration defined at the goal will be the system's **M**.

4.   THE OPTIMIZATION ALGORITHM

This problem is easily seen to be NP-hard (by reduction from the bin-packing problem (Korte and Vygen, 1991)). As such, this paper utilizes an approximating algorithm based on The Transportation Algorithm (Ore, 1992; and Strang, 1986) to determine system configuration. This algorithm is a well-documented optimization technique that has several similarities with the combinatorial issues of operator-function configuration. In summary, this algorithm determines a shipping network that maximizes the profits of a group of suppliers selling goods to a group of consumers such that the overall cost of shipping is minimized. Similarly, an optimal operator-function configuration is one where the number of functions receiving attention is maximized while the cognitive demand placed upon the human crew is minimized.

Despite the similarities, the Transportation Algorithm cannot perform workload optimization due to the following complications.

1. The Transportation Algorithm processes scalar quantities whereas cognitive capacity and cognitive loading are vector quantities.
2. A shipment matrix is composed of continuous, real number values whereas a configuration matrix contains discrete, Boolean values.

Complication (1). This situation requires modification to the vector components. As such, the cognitive capacity and loading vectors are collapsed into scalar components by selecting the smallest component from each $\bar{\Lambda}$ for operator representation and the largest component from each $\bar{\Gamma}$ for function representation. This reduction of dimensionality does affect the optimality of the configuration; however, the approximation drastically reduces the configuration space to improve the speed of computation.

Complication (2). This situation requires manipulation of continuous flow into discrete containers. To approximate an optimal configuration, the maximal flow component from each column of the configuration matrix is selected as the human-function match. Occasionally, selecting the maximal component can lead to cognitive overloading; however, if this case occurs, the work system varies the level of autonomy for any overloading functions and reprocesses the configuration.

It should be noted that these simplifications are only reasonable when the cognitive loading vectors are small relative to cognitive capacity vectors. When these vectors are on the same order of magnitude, the accuracy of approximation algorithm will degrade resulting in a systemic tendency to unnecessarily increase levels of functional autonomy.

5.   SIMULATIONS AND CONCLUSIONS

To demonstrate the capabilities of adaptive operator-robot work systems, a simulation was devised using the mathematic model and algorithm stated in this paper. The operator-robot work system was defined as a combat scenario involving two UAVs, five UGVs, and three operators. Each robot was given a unique set of subsystems (and functions) while operators were provided varying cognitive capacities to emulate variable levels of task training and operator skill.

A cost function in the form of $C = f(M_0, i, j) \cdot e^{g(p_i, q_i, Z_{ij})} \cdot (1 + \alpha A_{ij})$ was selected for simulation where C is the cost for pairing the ith operator to the jth function, M_0 is the preferred configuration, p is the functional priority, q is the system anticipation, Z is the task inertia, and A is the authority. The function $f(*)$ returns a constant that reflects the grouping assignment while $g(*)$ returns a composition of the priority, request time, and control inertia. This composition is placed into an exponential operator to give numerical importance to the mentioned parameters. Finally, α is selected to be significantly large such that when operator-robot parings are prohibited, C_{ij} becomes extremely costly thereby effectively prohibiting the potential pairing.

The results in Table 1 were generated using two methods: (1) the aforementioned procedure and (2) static assignments where operators where permitted to control pre-assigned functions until their cognitive threshold was reached. Autonomy alterations were requested based upon a Gaussian distribution of problem occurrence for each function (Figure 2). Ten trials were performed with a simulation length of thirty iterations. The results are expressed as the average number of functions (AF), the unassigned function per iteration (UF), the unassigned functions with high priority per iteration (UFHP), total autonomy switches (TAS), and total context switches (TCS).

Table 1. Simulation Results

AF	Method	UF	HPUF	TAS	TCS
8	Optimized	0.00	0.00	39.00	0
	Static	0.00	0.00	39.00	NA
31.3	Optimized	15.86	1.39	90.5	13
	Static	16.16	5.17	90.5	NA

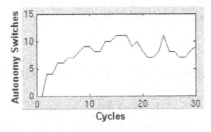

Figure 2. Autonomy level switching manifested simulations of the described scenario.

The first trial was a simple example demonstrating that systems with ample cognition have identical static and optimized results since operators were not cognitively overloaded in task management. The true benefits of the adaptive algorithm occur when the system becomes too complex for static assignments.   As in the second trial, the adaptive model (on average) supervised 0.3 more functions than static assignment while also allowing 3.78 more top-priority functions to be managed at each cycle. Clearly, if top-priority functions were absolutely critical for task completion, then the optimized procedure would be the preferred implementation.

Finally, adaptive operator-robot work systems show great potential as a supervising control mechanism. For example, simulation shows a 372% increase of high-priority function supervision over static assignment for each cycle. The cumulative improvement in functional management can also be quite significant when evaluated across missions having hundreds of thousands of cycles. Furthermore, system simplification (while not providing a guaranteed optimal solution) yields an approximation that can be estimated in a small polynomial time (i.e. the time complexity of the Transportation Problem) and is more amenable to the real-time control cycles demanded for mission critical operation in unstructured environments.

References

Anderson, J.R., and Lebiere, C. (1998). *The Atomic Components of Thought,* Mahwah, NJ: Erlbaum.

Barber, K.S., and Martin, C.E. (1999). Specification, Measurement, and Adjustment of Agent Autonomy: Theory and Implementation. *Autonomous Agents and Multi-Agent Systems.*

Fong, T., Thorpe, C., and Baur, C. (2001). Collaboration, Dialogue, and Human-Robot Interaction. In *Proceedings of 10th International Symposium of Robotics Research,* Lorne, Victoria, Australia.

Jordan, N. (1963). Allocation of functions between men and machines in automated systems. *Journal of Applied Psychology,* 47 (3).

Korte, B., and Vygen, J. (1991). *Combinatorial Optimization: Theory and Algorithms.* Springer. New York, NY.

Kortenkamp, D., Keirn-Schreckenghost, D., and Bonasso, R.P. (2000). Adjustable Control Autonomy for Manned Space Flight Systems. In *Proceedings of IEEE Aerospace Conference.*

Lebiere, C., Anderson, J. R., and Bothell, D. (2001). Multi-tasking and cognitive workload in an ACT-R model of a simplified air traffic control task. *In Proceedings of the 10th Conference on Computer Generated Forces and Behavior Representation.* Norfolk, Va.

Montemerlo, M. (2000). Uncertainty Based Sliding Autonomy. Doctorial Thesis Proposal. Carnegie Mellon University, Pittsburgh, PA.

Ore, O. (1962). *Theory Of Graphs.* American Mathematical Society, Providence.

Strang, G. (1986). *Introduction to Applied Mathematics,* Wellesley-Cambridge Press, Wellesley, MA.

USER INTERACTION WITH MULTI-ROBOT SYSTEMS

David Kortenkamp, Debra Schreckenghost and Cheryl Martin
NASA Johnson Space Center – ER2
Metrica Inc./TRACLabs
Houston TX 77058
kortenkamp@jsc.nasa.gov, ghost@ieee.org, cheryl.martin@jsc.nasa.gov

Abstract   There has been very little research on multiple human users interacting with multiple autonomous robots. In this paper we present some of the requirements of such user interaction. We present a prototype architecture for collaborative interaction. This architecture is put into the context of multiple space robots monitoring a space structure to assist human crew members.

Keywords: Multi-robot systems, human-robot interaction

1.   INTRODUCTION

Historically, human-robot interaction has focused on single humans interacting with single robots. For many deployed robot systems, there are multiple humans interacting with a single robot. Current research in multi-robot systems has often focused on robot-robot interaction and very rarely has it focused on human to multi-robot interaction. Even more rare is research focused on how multiple, cooperating humans interact with multiple, cooperating robots. In this paper we look at some of the key research issues in user interaction with multi-robot systems. An early architecture for handling these interactions is presented.

To provide context to the issues raised in this paper, we will use examples drawn from a specific class of NASA robots: mobile monitors. These are free-flying robots that have no manipulation capabilities. They can be used to perform inspection, monitoring and sensing tasks. Two examples of mobile monitors are currently under development at NASA. The first is the Personal Satellite Assistant (PSA), which is designed to be used inside of a space vehicle. The second is the Autonomous EVA Robotic Camera (AERCam), which is designed to be used outside of a space vehicle (see Figure 1). We will work

A.C. Schultz and L.E. Parker (eds.), Multi-Robot Systems: From Swarms to Intelligent Automata, 213-220.

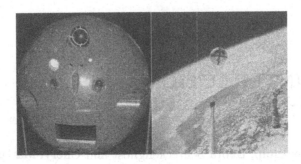

Figure 1.   Left: The Personal Satellite Assistant (PSA) robot. Right: The AERCam robot during a teleoperated space mission.

from a scenario in which there is one PSA, one AERCam and two crew members cooperating to perform a diagnosis task.

Our discussion will be based on a distributed interaction architecture currently being developed at NASA JSC. This architecture is meant to allow NASA crew members to interact with a wide range of autonomous systems (robots being just a subset). The architecture is still being defined and implemented, so this document simply reflects our initial direction. We will first present the distributed interaction architecture and then identify specific issues for robotic interaction.

2.   CONCEPTS FOR HUMAN-ROBOT INTERACTION

We have defined an operations concept for human-robot interaction that describes how humans and robots "should" be able to work together and identifies the challenges in achieving such interaction. We identify three main research areas: (1) monitoring and control of mostly autonomous robots; (2) managing the tasks of multiple, distributed agents (crew and robots); and (3) aiding distributed agents in exchanging information while operating remotely and asynchronously. Each of these research areas is described in this section.

2.1   Adjustable Autonomy

Our concept of robot operations includes robots that operate autonomously most of the time. Thus, an important user activity is maintaining situational awareness of autonomous robot activities and their effects. This supervisory task requires occasional, remote monitoring of the robots and user notification when interesting or unusual events occur or a need for manual action arises. Data summarization and visualization techniques are needed to support status-at-a-glance on the control situation across multiple, distributed robots. When situations arise that fully autonomous robots cannot address, it is necessary to support some level of user interaction and intervention with the robot. Our policy for such intervention is to provide for the interactive adjustment of autonomy (Kortenkamp et al., 2000; Barber et al., 2000). Techniques for such adjustment include reallocating autonomous tasks to the user, temporarily modifying autonomous procedures for unique situations, and overriding autonomous actions. Interactive adjustment of autonomy may also be needed when the user and robots work closely together to accomplish a shared mission objective. A key issue is determining whether/when/how to interrupt the robot.

2.2　Distributed Task Management

The coordination of user and robot activities will be a key component of the architecture. Coordination requires the capability to automatically track the activities of the users and the robots, to synchronize schedules when distributed robots and users share a common task objective, and to remind the users of pending tasks and task deadlines. When contingency situations with the robots arise, the user needs support in handling interruptions when they respond to the contingency. This includes assistance in managing multiple ongoing tasks, notification of task deadlines, and assistance in updating the activity plan.

2.3　Distributed User Communication

Distributing users and robots throughout a facility (such as space station) or outside the facility for extra vehicular activities (EVA) isolates users from indirect communication resulting from close proximity to the robots (e.g., they may not be able to observe an action) and can limit the information bandwidth available. Supporting operations where users and robots are occasionally out of communication further constrains the ability of each to maintain awareness of relevant operational changes. Finally, the increased use of autonomous robots takes the user out of the control loop, which can result in less awareness of ongoing robot actions. In such an environment, the user needs tools for effective information exchange in support of remote, asynchronous operations.

2.4　Robot Interaction Issues

The distributed interaction architecture applies equally to user interaction with robotic agents and with non-robotic software agents. In this section we look at a few issues that distinguish between "physical" agents (robots) and software agents.

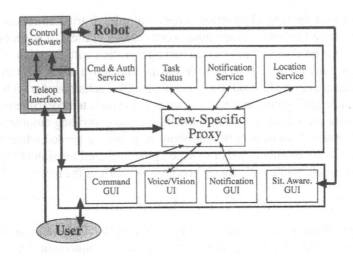

Figure 2.   The components comprising our distributed interaction architecture. This figure shows one user and one robot. Each additional user would have their own proxy, interface and services. Each additional robot would have its own control software and teleoperation interface.

- Teleoperation: There will need to be facilities for the user to take direct control of the robot, perform some task, and then turn control back over to the autonomous controller.

- Multi-modal input: The physical embodiment of the robot opens up new interaction possibilities, such as drawing a path to be followed on a graphical tablet. Also gestures can provide critical information.

- Physical interaction: Robots can interact physically with the user and the environment. These interactions require extra consideration for co-ordinating activities, possibly including models of physical processes.

Our initial implementation of the architecture has not addressed these robot interaction issues, but we expect to do so in the future.

3.   COMPONENTS OF THE ARCHITECTURE

We have developed a distributed interaction architecture design based on the operations concepts in the previous section and and a literature review (see Section 5). Components of this architecture that assist the user in robot interaction include a crew proxy for each user. A user interface mediates the interaction with these components. Each crew proxy utilizes the following *collaboration services*:

- A task status service that provides activity tracking and plan management capabilities for use by both the user and the robots.

- A location service that uses a variety of technologies to determine whether the user is online and to locate both user and robot relative to important features in environment (e.g., near the tank).

- A notification service that determines which user to notify of an event and how.

- A command and authorization service that determines which user has the authority to command a particular robot.

The services are self-explanatory, so we will concentrate on some other central components of the architecture next – the crew proxy and the user interface. Figure 2 shows all of these components.

3.1   Crew Proxy

Central to our approach for assisting the user in interacting with autonomous robots is providing crew proxies who represent and take action for the user. Each user has a proxy to represent his or her interests and concerns. From one perspective, this proxy stands in for the human agent in the interaction architecture. As a stand-in, it can interpret information from the robot or other agents, respond to requests from other agents and make decisions without bothering the human. From another perspective, this proxy augments human capabilities by enhancing human perception and actuation. As a capability augmentation, it can provide improved insight and enhanced ability to respond to robot situations.

The proxy provides functionality for a single user by coordinating collaboration services based on logical dependencies among these services. All user utilization of these services is mediated by the proxy software. The proxy also provides uniform access to information about its user that supports collaboration with other agents (robots or other users) in the architecture. It utilizes standardized knowledge models to delineate and represent collaborative information. Although our initial architecture only develops proxies for the crew, we believe that the proxy concept can be useful to a robot as well. This extension to the architecture permits crew proxies and robot proxies to collaborate without adversely affecting robot control.

3.2   User Interface

The bottom layer in Figure 2 contains components of a user interface. These components mediate the interaction between the user and the proxy, as well as providing a situation awareness interface that displays robot parameters (e.g., location, fuel, health, etc.). Teleoperation of the robot is achieved through a more direct connection with the robot control software, because there may be robot-specific requirements. Our goal is to provide a customizable interface for human-robot interaction.

4.   AN EXAMPLE SCENARIO

An example scenario is taken from a space domain and has two mobile monitors – one external to the spacecraft and one internal to the spacecraft and two crew members (users). In the scenario there is also an autonomous life support system. At the start of the scenario the two mobile monitors are engaged in routine inspection activities, with the status of those activities being monitored by the crew members. Event detection software for the life support system detects a possible gas leak. The crew members are alerted through their proxies in the way most appropriate for their current location and activity. While they work to reconfigure the system to bypass the leak, they dispatch (via their task planners) the mobile monitors to try to detect the leak – from both the inside and outside. The mobile monitors interface to the life support autonomous control system to narrow down their search. They both converge on the leak and notify the crew via the notification service and the proxies. A crew member uses its proxy (which ensures correct authorization) to take command (via teleoperation) of each mobile monitor in turn and fly in for closer looks. After rerouting gas around the leaks, the crew turns control of the mobile monitors back to their autonomous control systems and asks that they continue to monitor the leak site.

5.   LITERATURE REVIEW

Very little previous research has focused explicitly on multi-user interaction with multiple robotic agents. However, we can apply lessons from existing research on coordination and distributed collaboration among humans and software agents. We examined a number of implemented systems that helped inform the initial architecture design.

System characteristics, algorithms and interaction models that support different types of basic collaboration capabilities have been developed by previous research addressing human-agent interaction and mixed-initiative planning (i.e. COLLAGEN (Rich and Sidner, 1998) and TRIPS (Ferguson and Allen, 1998)) as well as overall coordination in distributed multi-agent systems (Jennings, 1996; Lesser, 1998). In particular, the proxy model of interaction between humans and software agents has been successfully demonstrated by the Electric Elves system (Chalupsky et al., 2001). In this system, proxy agents for each person in an organization perform organizational tasks for their users (e.g., monitoring the location of each user, keeping other users in the organization informed, and rescheduling meetings if one or more users is absent or unable to arrive on time). Previous research has also addressed other collaboration needs including the development of "advisable" agents that incorporate users' preferences about when to ask for permission or consultation for given behaviors (Myers and Morley, 2001).

Other previous research has developed integration infrastructure for multiagent systems: CoABS Grid (http://coabs.globalinfotek.com/), COUGAAR (http://www.cougaar.org), JADE (Bellifemine et al., 1999), KAoS (Bradshaw et al., 1997), RETSINA (Sycara et al., 2002), and the Open Agent Architecture (http://www.ai.sri.com/ oaa/whitepaper.html).

Knowledge representations and knowledge models supporting collaboration are critical to implementing a distributed interaction architecture. Various knowledge models have been employed by previous research to support collaboration and coordination. These models include task and activity models (Clancey et al., 1998), team and shared plan models (Kumar et al., 2000), resource and capability models (Chalupsky et al., 2001), user preference models (Myers and Morley, 2001), and roles, authority and organizational models (Bradshaw et al., 1997). These knowledge models and their representations together with the infrastructure support provided by multi-agent development platforms and distributed computing technologies provide a foundation for the implementation of the architecture. By leveraging this previous work concerning software agents, we can make rapid progress toward building a system to support multi-user/multi-robot interaction.

Acknowledgments

This work is supported by several NASA research grants. Including a grant from the NASA Intelligent Systems Program, Human Centered Computing and a grant from the NASA Engineering Complex Systems program. Discussions with Carroll Thronesbery (NASA JSC/SKE Incorporated), Pete Bonasso (NASA JSC/Metrica Inc.) and Tod Milam (NASA JSC/Metrica Inc.) contributed to the architecture.

References

Barber, S., Goel, A., and Martin, C. (2000). Dynamic adaptive autonomy in multi-agent systems. *Journal of Experimental and Theoretical Artificial Intelligence*, 12(2).

Bellifemine, F., Poggi, A., and Rimassa, G. (1999). JADE - a FIPA-Compliant agent framework. In *Proceedings of the Fourth International Conference and Exhibition on the Practical Application of Intelligent Agents and Multi-Agents (PAAM'99)*.

Bradshaw, J. M., Dutfield, S., Benoit, P., and Woolley, J. (1997). KAoS: Toward an industrial-strength generic agent architecture. In Bradshaw, J. M., editor, *Software Agents*. AAAI/MIT Press, Cambridge MA.

Chalupsky, H., Gil, Y., Knoblock, C. A., Lerman, K., Oh, J., Pyandath, D. V., Russ, T. A., and Tambe, M. (2001). Electric elves: Applying agent technology to support human organizations. In *Proceedings of the Innovative Applications of Artificial Intelligence*.

Clancey, W. J., Sachs, P., Sierhuis, M., and van Hoof, R. (1998). Brahms: Simulating practice for work systems design. *International Journal of Human-Computer Studies*, 49.

Ferguson, G. and Allen, J. (1998). TRIPS: an integrated intelligent problem-solving assistant. In *Proceedings of the Fifteenth National Conference on Artificial Intelligence (AAAI-98)*.

Jennings, N. R. (1996). Coordination techniques for distributed artificial intelligence. In O'Hare, G. M. P. and Jennings, N. R., editors, *Foundations of Distributed Artificial Intelligence, Sixth-Generation Computer Technology Series,*. John Wiley and Sons, New York.

Kortenkamp, D., Keirn-Schreckenghost, D., and Bonasso, R. P. (2000). Adjustable control autonomy for manned space flight. In *IEEE Aerospace Conference*.

Kumar, S., Cohen, P. R., and Levesque, H. J. (2000). The adaptive agent architecture: Achieving fault-tolerance using persistent broker teams. In *Proceedings of the International Conference on Multi-Agent Systems*.

Lesser, V. R. (1998). Reflections on the nature of multi-agent coordination and its implications for an agent architecture. In *Autonomous Agents and Multi-Agent Systems (AAMAS-98)*.

Myers, K. L. and Morley, D. N. (2001). Directing agent communities: An initial framework. In *Proceedings of the IJCAI-2001 Workshop on Autonomy, Delegation, and Control: Interacting with Autonomous Agents*.

Rich, C. and Sidner, C. L. (1998). COLLAGEN: a collaboration manager for software interface agents. *User Modeling and User-Adapted Interaction*, 8(3-4).

Sycara, K., Paolucci, M., van Velsen, M., and Giampapa, J. (2002). The RETSINA MAS infrastructure. In *Autonomous Agents and Multi-Agent Systems*.

HUMAN-ROBOT INTERACTIONS IN ROBOT-ASSISTED URBAN SEARCH AND RESCUE

Robin Murphy and Jenn Casper
Computer Science and Engineering
University of South Florida
murphy@csee.usf.edu

Abstract   Robot-assisted urban search and rescue is an exemplar of a domain where humans must interact with robots and the information they produce, providing capabilities that do exist in traditional methods. It is also a multi-agent domain, where a human might interact with multiple heterogeneous agents. The paper presents data collected in field exercises with subject matter experts (Florida Task Force 3, Hillsborough County Fire Rescue Department, and Fire Department New York/Federal Emergency Management Agency at the World Trade Center). Analysis suggests that execution errors (e.g., controlling the robots) are more prevalent than errors of intention (e.g., where to use robots), the lack of multi-modal interfaces interferes with task completion, many activities are prototypical and could be encapsulated as autonomous behaviors or schemas, and topological mental models are acceptable and perhaps preferable over metric.

A.C. Schultz and L.E. Parker (eds.), Multi-Robot Systems: From Swarms to Intelligent Automata, 221.
© 2002 *Kluwer Academic Publishers. Printed in the Netherlands.*

USABILITY ISSUES FOR DESIGNING MULTI-ROBOT MISSIONS

Ronald C. Arkin
College of Computing
Georgia Institute of Technology
Atlanta, Georgia 30332-0280

Abstract   As part of our research for the Defense Advanced Research Projects Agency (DARPA) Tactical Mobile Robotics Program, we have conducted a series of usability studies evaluating the ability of users to design missions (both single and multi-robot) for various scenarios (e.g., airport incursion, biohazard survey and detection). This talk presents the results of this study involving dozens of human subjects working with Georgia Tech's MissionLab mission specification system, and provides insights into the issues surrounding pre-mission planning for humans tasking teams of mobile robots.

A.C. Schultz and L.E. Parker (eds.), Multi-Robot Systems: From Swarms to Intelligent Automata, 223.

PERCEPTION-BASED NAVIGATION FOR MOBILE ROBOTS

K. Kawamura, D. M. Wilkes, A.B. Koku, and T. Keskinpala
Intelligent Robotics Laboratory, Vanderbilt University
Nashville, TN, 37235

Abstract: Ongoing research on multi-robot collaborative navigation conducted in the Intelligent Robotics Lab at Vanderbilt University is introduced. Range-free egocentric representations for navigation have been previously developed for a single robot. We refer to this method as *perception-based navigation* (PBN). This paper will show that the concept of PBN can be extended from a single robot to a group of robots to achieve collaborative navigation. One robot attempts to reach a target region by sharing its sensory information with the knowledge of another. PBN approach separates sensing and navigation from planning. Our assumption is that one robot is lost or new in an environment, which is known to another robot. The new robot asks for assistance by presenting its egocentric world perceptions. We show the PBN idea is expandable to handle collaborative efforts to achieve a goal.

Keywords: Navigation, perception, sensory egosphere, landmark egosphere

1.   INTRODUCTION

The goal of the work reported in this paper is to develop a new range-free geometrically-based navigation algorithm to enable a robot to navigate in an unknown environment by using knowledge from another robot about that region. By range-free, we mean no distance information to the landmarks around the robot is explicitly sought. By geometrically-based, we mean that geometric rather than metric information is required for navigation.

The idea is to have each robot use given via-points to define *via-regions*, which the robot can navigate reactively. Each robot senses the world.

A.C. Schultz and L.E. Parker (eds.), Multi-Robot Systems: From Swarms to Intelligent Automata, 225-234.

Independent sensory processing modules register their results on a *Sensory EgoSphere* (SES), an egocentric 2D spherical map of the robot's current surroundings. Each robot has a sparse, metrically imprecise map of the territory in which it operates. When given a via-point by the operator, a robot projects onto a new EgoSphere, called a *Landmark EgoSphere* (LES). The landmarks are what it may sense from the vicinity of the via-point. The robot reaches its next via-region by heading in the direction of a landmark visible from both the current location and the via-region while comparing the current contents of its SES to the LES of the target via-region.

Our approach is inspired by the qualitative navigation method of Levitt and Lawton 0 and Dai and Lawton 0. Perception-Based Navigation (PBN) addresses egocentric and allocentric representations along with the concepts of short- and long-term memories to provide a range-free geometric solution to navigation. The methodology described for a single robot can be expanded to a group of robots. In this paper, we will define the SES and the LES, explain their use in terms of short- and long-term memory structures, and describe the navigation algorithm. We will present an example of two robots in which one uses its own sensory input plus the terrain knowledge of another to navigate towards a target. We have designed a behavior-based architecture for the robots that runs under the Intelligent Machine Architecture (IMA), an agent based software system that permits concurrent execution of software objects on a network of computers (Pack, 1999; Pack, Wilkes, and Kawamura, 1997).

2.   EGOSPHERE AND MEMORY MODELS

Albus proposed the EgoSphere in 1991. He envisioned it as a dense map of the visual world, a virtual spherical shell surrounding the robot onto which a snapshot of the world was projected. Our definition and use of the EgoSphere differs from that of Albus. We call it the Sensory EgoSphere (SES) (Kawamura, *et al.*, 2001a) and define it as a 2D spherical data structure, centered on the coordinate frame of the robot that is spatially indexed by azimuth and elevation. Its implicit topological structure is that of a geodesic dome, in which each node is a pointer to a distinct data structure.

During navigation, targets, landmarks, and external and internal events stimulate the sensors. Upon receiving a stimulus, the associated sensory processing module writes its output data (including the time of detection) to the SES at the node that is closest to the direction from which the stimulus arrived as illustrated in Figure 1.

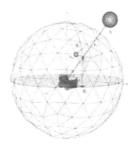

Figure 1. Mapping an object to EgoSphere. Object S lies along r_s and has the coordinates θ_s, Φ_s. Ns is the vertex closest to θ_s, Φ_s.

Often in practice SES is not a complete geodesic dome, instead, it is restricted to only those vertices that fall within the directional sensory field of the robot. In this paper, the perceived landmarks are projected onto the equator similar to ship navigation where distant landmarks are located on the horizon. Since the PBN algorithm uses only the relative angles of objects with respect to the robot's heading, the elevation of perceived objects is neglected.

Our robot navigation system comprises local and global paradigms. Local navigation is reactive and based solely on information within the immediate sensing region of the robot. Global navigation uses information from beyond the limits of local sensors (the *sensory horizon*) and is deliberative. This division implicitly organizes the robot's memory into short- and long-term components. Long-term memory (LTM) is, in part, spatially organized to support global navigation, and includes a map of landmarks that should be expected by the robot, and objects that it has detected during its current and previous forays.

Short-term memory (STM) for reactive navigation is provided by the SES. On its vertices, SES may contain links to data structures in the LTM of the robot. STM is robo-centric, sparse, and has limited duration determined by factors such as object properties, informational uncertainties, among others.

At any specific location in the environment, the sensory horizon defines the region in which the robot can sense. Only those objects within the region have the potential to be sensed and stored on the SES. During the navigation, the robot updates its STM periodically, in effect creating instances of SES structures at discrete locations. Each SES instance is a snapshot of the environment at a specific space-time location. These instances form a chain of SES representations (Figure), which define a topological map – a collection of nodes and connecting arcs. At the end of navigation, a series of SES regions are stored in the STM.

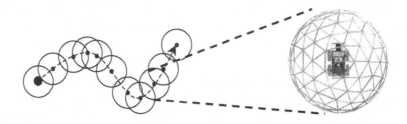

Figure 2. Chain of SES regions stored in the short-term memory each representing a geodesic dome around the robot 0.

The robot's long-term memory contains global layout map information. The structural layout and major landmarks are indicated on the rough *a priori* map that is loaded into the robot. This map may be subsequently modified by the topological data stored in the STM. At any given position, landmarks presumed to be surrounding the robot are represented by a LTM structure called the Landmark EgoSphere (LES). The LES is a robo-centric representation of environmental features expected at the current position that is extracted from the long-term memory. Within our navigational scheme, the directional information provided by the LES is of prime importance to the robot control system. The robot also uses a *working memory* module that holds descriptors of key locations (via-regions) in terms of LES representations that indicate transition points for navigation. Detailed information on these can be found in (Kawamura, *et al.*, 2001b).

3.   SES- AND LES-BASED NAVIGATION

One of the main factors in the different approaches to solving navigation problems is in the way objects in the environment are represented within the robot, i.e., the representation problem. With this in mind, we have reflected upon how humans address similar problems.

Consider the case where Person A needs to reach an office in a certain building. Not knowing how they can reach this building, he/she asks for directions from Person B who provides a sketch describing, qualitatively, how to reach this building from the current location. This example illustrates the essential components of what we want to achieve by PBN. In this analogy, the interaction among Persons A and B represents the *information sharing.* The sketch assists the person towards the target location is a key point in this approach. It is neither metrically precise nor accurate; however, it is descriptive enough so that the person can follow certain landmarks on the sketch to reach the target. Very often sketches do not pinpoint the target, but direct the person close enough to the target region. The inaccuracies of the sketch are compensated by the perception of the actual scene and assumed reasoning capacity of the person. The spatial definitions are also very relaxed, i.e. *"go to front of the library"*. The front of the library is a region rather than a point and yet descriptive enough for humans.

This type of human behavior led us to the PBN approach. First, we try to relax the terms of localization. Rather than using point-wise localization and navigation, we try to create an understanding of *"regions"* and use this concept in navigation and localization. Second, we do not try to be very precise until it is necessary. When the above example is considered, a person does not seek absolute positioning until it is needed. Being in the region, the front of the library is enough. More precise navigation tasks such as passing through a narrow doorway or picking up a certain object may require more precision with respect to position. Therefore, we categorize reaching a certain region as an imprecise action. Performing an action within this region may require more precision in terms of localizing the robot.

3.1   Perception and Representation

In this paper, perception is defined as the ability of the robot to derive useful abstractions based on raw sensory readings. Currently, our robots can distinguish a limited number of objects based on saliency of colors. Recently, we developed a visual attention algorithm based on defining the concept of a "bright" color in the HSV color space. Preliminary tests on the actual robot are promising. This new perception scheme is more robust to lighting changes and uses fewer parameters to define interesting colors for landmarks.

When one tries to remember a place or a scene, geometric relationships between objects in the environment are more naturally remembered than the metric relationships between them. Recent studies suggest that humans favor angular information over distance information while learning places and localizing themselves 0. There are also other biological findings that suggest that angular representations are used to store spatial information as well as navigation. Bee navigation is such an example (Cartwright and Collett, 1983; Menzel, *et al.*, 2000). This is consistent with what we are trying to do while achieving range-free, region-based PBN. A *region* is described by the landmarks that are visible or are expected to be visible to the robot from within that region.

A priori information provided to the robot will be saved in a LTM structure (i.e., the rough initial map), which accepts and stores imprecise qualitative information on a 2D plane. From this memory structure, for any given point in the map the robot can extract a rough angular distribution of the objects around that point. This angular distribution is the previously described LES. Similarly, the current perception of the robot is also represented by the angular distribution of the objects around the robot in the form of SES. By comparing the SES with LES, the robot is able to navigate to a target region.

3.2   Perception-Based Navigation

Achieving range-free PBN is our goal. We have developed a simple algorithm that moves the robot towards a target location based on the current perception. Figure 3 illustrates key concepts in PBN. The sketchy map is stored in the LTM along with the icons for known objects. This long-term memory representation is used to extract LES representations to assist the robot's navigation.

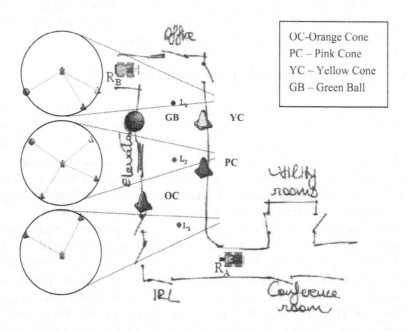

Figure 3. Key concepts in PBN: Long-term memory, SES and LES.

Since range information is not explicitly sought, the robot represents its current perception by using the angular separation between perceived objects. Currently, our robot's depth of vision is limited to 5-6 meters based on the objects used. Figure 3 shows three angular representations based on three points L_1, L_2 and L_3. These are egocentric, range-free representations of the robot's actual or expected surrounding. If the robot is actually there and creates this representation by sensing the world, then it is represented by SES. However, if the robot uses its long-term memory to see what the world is expected to look like once it reaches that point, than it is a LES representation. Therefore, any path plan, including the via-regions and the target, is expressed as a collection of LES regions placed in the working memory.

The navigation algorithm simply works as follows. The robot senses the world and creates an SES, which is compared to the LES marked as the immediate target. This comparison creates a vector by moving in the direction, which improves the similarity between SES and LES. Therefore, this vector is converted to motion and while moving along this direction the robot iteratively creates SESs to update its heading until the difference between the latest SES and target LES is below a certain error threshold.

Thus, the basic navigation behavior only depends on two range-free representations, which makes this algorithm extensible. The LES representation may come from the long-term memory, another robot or human commander. Therefore, this method can be used in collaborative navigation as it is presented in the following section. It is also possible to turn the table around and use this navigation method for localization purposes over the long-term memory. If the currently sensed SES is treated as LES and the long-term memory acts as the real world, starting from the last known location of the robot, a simulative navigation can be carried out over the long-term memory to yield localization.

4. SCENARIO: MULTI-ROBOT NAVIGATION BASED ON PBN

Robots R_A and R_B are located as shown in Figure 3. R_A does not know this environment, but R_B does. The goal for R_A is to reach R_B using knowledge from R_B. R_A shares its SES with R_B. R_B then determines the region that R_A is in and sends a heading and the next LES for R_A to reach using PBN. This is analogous to the situation where someone who is lost speaks to another person by cell phone, describing what they see, and obtaining directions to a certain point.

In the example illustrated in Figure 3, R_A initially sees only the orange cone. At this point, it needs a target LES to navigate. R_A passes its object-angle representation pair to R_B. Based on this, R_B returns the first target LES at L_1. The LES at L_1 contains the orange cone. There should be at least one common object between the current SES and the target LES. It should also be noted that LES depth is practical. That is, at L_1, R_A may be sensing more objects than there are in the target LES depending on its sensory horizon, but what matters and what is transferred to R_A are the expected object-angle distribution at the point L_1. After getting the target LES at L_1, R_A uses PBN to reach L_1. Upon reaching L_1, R_A passes its current object-angle pair, and asks for another target. Then, R_B passes the target LES at L_2. Once R_A reaches L_2, R_B passes the target LES at L_3. The information sharing and navigation continues until both robots sense each other.

In this scenario, the goal is achieved by sharing minimal information between robots. Only the object-angle representation pairs are passed back and forth between the robots, thus minimizing communication time.

To date, PBN has proven to work properly for a single robot where the robot uses its own memory. Preliminary work on achieving collaborative navigation has been implemented successfully.

5.   CONCLUSIONS

This paper described research work in progress to develop a range-free and geometrically-based navigation algorithm for mobile robots. The robot control architecture used is multi-agent based, originally developed for a stationary robot (Kawamura, *et al.*, 2000). The SES concept has been extended to include a graphical display agent for the GUI for another DARPA project (Kawamura, *et al.*, 2001a). Efforts are under way to extend the results reported here to include multiple landmarks of similar kinds.

Acknowledgments

This research has been partially funded by the DARPA Mobile Autonomous Robotics Systems (MARS) program (Grant #DASG60-99-1-0005).

Appendix

In this paper, two robots are involved. Robot R_A is a Pioneer 2-AT robot equipped with a $360°$ vision system composed of 7 cameras and an RS232-controlled video multiplexer interfaced to a PC via a USB digitizer (Figure 4).

Robot R_B is an ATRV junior with a pan-tilt camera head. An extra camera is added to the pan-tilt head pointing in the opposite direction to provide 360° vision. All computers are connected to the LAN via wireless network cards. No other sensory information such as odometry or GPS is used.

a. Pioneer 2-AT robot with added hardware. b. Camera array installed on Pioneer 2-AT

Figure 4. Test Bed: Pioneer 2-AT robot

References

Albus, J.S. Outline for a theory of intelligence. (1991). *IEEE Transactions on Systems, Man and Cybernetics*, 21: 473-509.

Cartwright. B.A., and Collett, T.S. (1983). Landmark learning in bees. *Journal of Comparative Physiology A* 151: 521-543.

Dai, D., and Lawton, D. T. (1993). Range-free qualitative navigation. In *Proceedings of the IEEE International Conference on Robotics and Automation*, pages 783-790.

Healy, S. (1998). *Spatial Representation in Animals*, Oxford University Press, pages 6-8.

Kawamura K., Peters II, R.A., Johnson, C., Nilas, P., and Thongchai, S. (2001). Supervisory Control of Mobile Robots using Sensory EgoSphere. In *Proceedings of IEEE International Symposium on Computational Intelligence in Robotics and Automation*, pages 523-529, Banff, Alberta, Canada.

Kawamura, K., Peters II, R.A., Wilkes, D.M., Koku, A.B., and A. Sekmen. (2001). Toward Perception-Based Navigation Using EgoSphere. In *SPIE 2001*, Boston, MA.

Kawamura, K., Peters II, R.A., Wilkes, D.M., Alford, W.A., and Rogers, T. E. (2000). ISAC: Foundations in Human-Humanoid Interaction. *IEEE Intelligent Systems*, (July/August):38-45.

Levitt, T. S., and Lawton, D. T. (1990). Qualitative navigation for mobile robots. *Artificial Intelligence*, 44:305-360.

Menzel, R., Brandt, R., Gumbert, A., Komishchke, B., and Kunze, J. (2000). Two spatial memories for honeybee navigation. In *Proceedings of the Royal Society of London Series*

B--Biological Sciences, 267, pages 961-968.Pack, R.T. (1999). IMA: The Intelligent Machine Architecture. Ph.D. Dissertation, Vanderbilt University.

Pack, R.T., Wilkes, D. M., and K. Kawamura. (1997). A Software Architecture for Integrated Service Robot Development. In *Proceedings of IEEE Conf. On Systems, Man, and Cybernetics*, pages 3774-3779, Orlando, Florida, USA.

Author Index